Bioenergy and Biofuels

Bioenergy and Biofuels

Editors

Karol Tucki
Olga Orynycz

MDPI • Basel • Beijing • Wuhan • Barcelona • Belgrade • Manchester • Tokyo • Cluj • Tianjin

Editors
Karol Tucki
Warsaw University of Life
Sciences
Poland

Olga Orynycz
Bialystok University of Technology
Poland

Editorial Office
MDPI
St. Alban-Anlage 66
4052 Basel, Switzerland

This is a reprint of articles from the Special Issue published online in the open access journal *Sustainability* (ISSN 2071-1050) (available at: https://www.mdpi.com/journal/sustainability/special_issues/Bioenergy_Biofuels).

For citation purposes, cite each article independently as indicated on the article page online and as indicated below:

LastName, A.A.; LastName, B.B.; LastName, C.C. Article Title. *Journal Name* **Year**, *Volume Number*, Page Range.

ISBN 978-3-0365-2051-3 (Hbk)
ISBN 978-3-0365-2052-0 (PDF)

Contents

About the Editors

Karol Tucki Institute of Mechanical Engineering, Faculty of Production Engineering Warsaw University of Life Sciences, Nowoursynowska Street 164, 02-787 Warsaw, Poland Scientific interests: engines; fuels; turbines; energy engineering; environment; sustainability. Education: Ph.D. degree in Building and Exploitation Machines Science (Gdansk Univ. Tech. 2006), M.Sc. Eng. in Mechanic and Construction of Machines Science (Olsztyn University of Warmia and Mazury 2002–2004), Eng. in Mechanical Engineering Science (Elblag The State University of Applied Sciences 1998–2002). Professional experience: Warsaw University of Life Sciences (from 2012), GE Company (2008–2012), Engineering Design Center—Warsaw Institute of Aviation (2007–2008).

Olga Orynycz Assistant Professor at Bialystok University of Technology, Dept. Production Management. The Board Secretary of Podlaskie Chapter of Polish Association of Production Management. The representative of Bialystok University of Technology in the Council of Business Cluster under the name Partnership for Knowledge and innovation in Production Engineering. The vice-chairman for promotion at The Board of School Club AZS (Academic Sports Association) at Bialystok University of Technology. Scientific Mentor for the area of Production, as the representative of the Department of Engineering Management (2015–2017). The Author of present proposal performs research in the production engineering; renewable energy technologies; mathematical modelling; biofuels; production management.

sustainability

Editorial

Bioenergy and Biofuels

Karol Tucki [1,*] and Olga Orynycz [2,*]

[1] Department of Production Engineering, Institute of Mechanical Engineering, Warsaw University of Life Sciences, Nowoursynowska Street 164, 02-787 Warsaw, Poland
[2] Department of Production Management, Bialystok University of Technology, Wiejska Street 45A, 15-351 Bialystok, Poland
* Correspondence: karol_tucki@sggw.edu.pl (K.T.); o.orynycz@pb.edu.pl (O.O.); Tel.: +48-593-4578 (K.T.); +48-746-9840 (O.O.)

Citation: Tucki, K.; Orynycz, O. Bioenergy and Biofuels. *Sustainability* **2021**, *13*, 9972. https://doi.org/10.3390/su13179972

Received: 30 August 2021
Accepted: 5 September 2021
Published: 6 September 2021

Publisher's Note: MDPI stays neutral with regard to jurisdictional claims in published maps and institutional affiliations.

Technology is setting strategic directions for change in the economy and industry. Among the global forces that we are currently observing and that will have the greatest impact on the shape of the world economy in the coming years are Industry 4.0, the circular economy model, sustainable finance, the talent market, the "silver economy" and electromobility. These affect and will affect for many years the course of many processes such as production, consumption, investments and environmental protection activities. The health of the fuel and energy sector affects economic development around the world. New technologies in the energy sector and the management of its development together with the dynamically changing environment, as well as care for sustainable development and energy security, make the power industry and the automotive industry the most important sectors of the economy, and their dynamic development has been observed for many years. Both conventional energy and motorization are the largest source of gas emissions into the atmosphere. The financial burden for exceeding permitted emissions, as well as public awareness, forces this to change.

The objective adopted by the European Union to reduce greenhouse gas emissions by 55% by 2030 and achieve full climate neutrality by 2050 means that the market for alternative energy sources must be developed dynamically. The regular increase in the share of renewables in the global energy mix for many years indicates that many public and private institutions are making efforts to decarbonize the economy. As a result of intensified energy transition efforts and the introduction of the so-called European Green Deal, entirely new ecosystems and new technologies are emerging. Various alternatives are being explored to reconcile economic growth with care for the environment. Product Life Cycle Assessment and the Closed Circuit Economy are helping to unleash innovation and technological progress. Ambitious climate neutrality goals cannot be achieved without alternative, low or zero carbon, energy and fuel technologies. The intensity of changes taking place in the fuel and energy sector, both at the regulatory and technological level, forces the research community and scientists to carry out new, increasingly interesting research. The pressures of energy transition present many challenges but also opportunities for both economic, technological and scientific development.

It should be stressed that an important factor mobilising the search for new technologies, especially in the energy and transport sectors, is the progressive climate change closely linked to greenhouse gas emissions.

The changes taking place are contributing to a dynamic transformation not only of activities based on conventional fuels, but also to the development of renewable energy sources. Many of the current moves towards decarbonization and climate neutrality present science and business with previously unknown and unique challenges, opportunities for transformation and growth. Decarbonisation means facing up to big changes. The way we obtain, use, consume and generally think about energy and raw materials needs to be revised. The current level of technological development and, above all, significant decreases in the costs of introducing the latest solutions facilitate the adoption of effective decarbonisation strategies. Given the scale of the challenge to achieve full climate neutrality, there is a need for commitment from the legislative side, from the scientific side, from business and from public institutions implementing the solutions developed using all current and future technologies and resources.

The Special Issue "Bioenergy and Biofuels" of the journal Sustainability was dedicated to the publication of works on obtaining energy from biological sources. Obviously, bio-based biomass contains mainly carbon and hydrogen and can be converted into various types of fuel or burned directly to provide heat. From the composition of biomass, it can be easily deduced that its combustion mainly causes the emission of carbon dioxide and water. Carbon dioxide from biomass is assumed to have been absorbed from the atmosphere during plant growth and will be reabsorbed. Therefore, it is not a source of climate warming, and it only temporarily increases the concentration of carbon dioxide in the atmosphere. This situation is the main reason for the use of plant biomass for energy purposes. Due to the variety of applications, there are many technologies for obtaining energy from biomass. New technologies for obtaining as well as technologies for converting bio-based fuels into various forms of energy may also emerge. The use of renewable energy sources is governed by a number of legal provisions on various aspects of the conversion of biomass into fuels, the use of waste biomass, etc. All these aspects are reflected in five published articles.

The first paper presents an analysis of the implementation of the Paris Agreement and recommendations for the reduction of greenhouse gas emissions in the EU in relation to the so-called countries of the Visegrad Group (V4), i.e., Poland, the Czech Republic, Hungary and Slovakia. It analyses the structures of energy production, its consumption over the years, and analyses the measures taken to improve energy efficiency.

The next article discusses the potential of wood fuel in the Swiss economy. It was stressed that developing an energy transition and decarbonisation strategy requires consideration of the different types of wooden biomass. The forecast of changes in the theoretical and sustainable potential of wood fuel from various wood resources was presented indicating perspectives of growth of sustainable potential of wood based fuels in the not distant future. However, the development of a circular economy and organization of the wood industry may play an important role as a condition assuring this growth.

The next work is devoted to the study of the behaviour of biodiesel produced from palm oil and its blends with petroleum diesel when burned in a test engine. It has been shown that biodiesel has a higher cetane number than classic fossil fuel and is well suited for engine propulsion, both in a pure state and in blends.

The research on the influence of the character of row wooden material as well as the conditions of the agglomeration process of wood briquettes on their quality and mechanical properties will be the subject of another publication. Here, a significant influence of all the specified factors has been demonstrated and the optimal technological parameters have been selected to obtain high quality briquettes.

The last article in this issue is a literature review on the use of microalgae farming as a source of biomass for energy purposes. Despite the high potential of microalgae as a raw material for the production of biofuels, a number of problems in breeding and sourcing make it difficult to commercialize. The paper presents a number of solutions concerning both biological aspects and the technology of obtaining oily substances and their further processing. It should be mentioned that production of biofuels from organisms living in water is important from the viewpoint of competition between industrial and food products observed in land-based agriculture.

The Guest Editors of the Issue would like to express their thanks to the authors of the articles for the high quality of the presented works.

Funding: This research received no external funding.

Institutional Review Board Statement: Not applicable.

List of Contribution

1. Tucki, K.; Krzywonos, M.; Orynycz, O.; Kupczyk, A.; Bączyk, A.; Wielewska, I. Analysis of the Possibility of Fulfilling the Paris Agreement by the Visegrad Group Countries.
2. Erni, M.; Burg, V.; Bont, L.; Thees, O.; Ferretti, M.; Stadelmann, G.; Schweier, J. Current (2020) and Long-Term (2035 and 2050) Sustainable Potentials of Wood Fuel in Switzerland.
3. Nguyen, V.H.; Duong, M.Q.; Nguyen, K.T.; Pham, T.V.; Pham, P.X. An Extensive Analysis of Biodiesel Blend Combustion Characteristics under a Wide-Range of Thermal Conditions of a Cooperative Fuel Research Engine.
4. Nurek, T.; Gendek, A.; Roman, K.; Dąbrowska, M. The Impact of Fractional Composition on the Mechanical Properties of Agglomerated Logging Residues.
5. Shokravi, Z.; Shokravi, H.; Chyuan, O.H.; Lau, W.J.; Koloor, S.S.R.; Petrů, M.; Ismail, A.F. Improving 'Lipid Productivity' in Microalgae by Bilateral Enhancement of Biomass and Lipid Contents: A Review.

sustainability

Article

Analysis of the Possibility of Fulfilling the Paris Agreement by the Visegrad Group Countries

Karol Tucki [1,*], Małgorzata Krzywonos [2], Olga Orynycz [3,*], Adam Kupczyk [1], Anna Bączyk [1] and Izabela Wielewska [4]

[1] Department of Production Engineering, Institute of Mechanical Engineering, Warsaw University of Life Sciences, Nowoursynowska Street 164, 02-787 Warsaw, Poland; adam_kupczyk@sggw.edu.pl (A.K.); anna_baczyk@sggw.edu.pl (A.B.)

[2] Department of Process Management, Wrocław University of Economics and Business, Komandorska Street 118/120, 53-345 Wrocław, Poland; malgorzata.krzywonos@ue.wroc.pl

[3] Department of Production Management, Bialystok University of Technology, Wiejska Street 45A, 15-351 Bialystok, Poland

[4] Department of Agronomy, UTP University of Science and Technology in Bydgoszcz, Fordońska Street 430, 85-790 Bydgoszcz, Poland; wielewska@utp.edu.pl

* Correspondence: karol_tucki@sggw.edu.pl (K.T.); o.orynycz@pb.edu.pl (O.O.); Tel.: +48-593-4578 (K.T.); Tel.: +48-746-9840 (O.O.)

Abstract: The aim of this study was to analyse the feasibility of implementing the Paris Agreement and the provisions regarding the goals of reducing greenhouse gas emissions in the EU through Poland, the Czech Republic, Hungary and Slovakia, i.e., the so-called Visegrad Group States (V4). The basis of the study was an in-depth analysis of the energy policies of the V4 countries, an analysis of energy generation structures, its consumption over the years, and an analysis of measures taken to improve energy efficiency. The analysis was performed as a function of the adopted targets for reducing CO_2 emissions by 2020, with a prospect for 2030 and 2050. In all the analysed countries, the energy and heat production sectors were responsible for the highest carbon dioxide emissions. Among the analyzed countries, only Poland failed to meet its commitments regarding the level of greenhouse gas (GHG) reductions adopted by 2020. The achievement of further goals in this area, despite the planned investments and undertaken actions, is also at risk

Keywords: international agreements; climate change; Paris Agreement; Visegrad Group; greenhouse gas emissions

Citation: Tucki, K.; Krzywonos, M.; Orynycz, O.; Kupczyk, A.; Bączyk, A.; Wielewska, I. Analysis of the Possibility of Fulfilling the Paris Agreement by the Visegrad Group Countries. *Sustainability* **2021**, *13*, 8826. https://doi.org/10.3390/su13168826

Academic Editors: Tomonobu Senjyu, Brantley Liddle and Marc A. Rosen

Received: 9 June 2021
Accepted: 5 August 2021
Published: 6 August 2021

1. Introduction

After setting the Paris Agreement, the term "climate policy" undeniably has a new dimension [1,2]. The agreement gives an innovative approach to climate protection issues with ambitious goals and creates the obligations assumed by individual states [3,4]. It is a comprehensive strategy that combines environmental issues with all sectors of the economy, main players in every country [5,6]; this is the problem faced in the modern world [7,8]. Therefore, integrated cooperation in energy, transport, construction, public administration, and environmental policy in a given country is required for global warming mitigation [9,10]. Closer cooperation between the various parties at the international level is necessary [11]. The European Commission recommends exchanging scientific knowledge on adaptation to climate change and information on behavior and strategy [12,13]. Combating climate change should remain on the top of the political lists of priorities of relevant international forums [14,15]. Another crucial thing is that after signing Paris Agreement, the parties must have the reference point to calculate their emissions, and in most cases, it is 1990.

The Visegrad Group (V4) is one example of strengthening international cooperation in Central Europe recommended by the European Commission to implement the Paris

5

Agreement [16,17]. The V4 community is a regional form of cooperation between Poland, the Czech Republic, Slovakia, and Hungary [18,19]. Besides a close neighbourhood and geopolitical features, the members of the V4 share a common history, culture, traditions and religious and intellectual values that they wish to preserve and strengthen [20,21]. While being an important part of regional cooperation, the group constitutes an element that strengthens the international position and is conducive to the social and economic growth of the four states.

The energy crisis of 2009 showed the gas dependence of the Visegrad Group member states [22]. Their domestic production is significantly less than what is needed to satisfy sufficient supplies to consumers, and imports of gas and other forms of energy depend primarily on Russia. While Poland's foreign policy sought to ensure that Poland avoided any cooperation with Russia, Hungary took steps to strengthen ties through economic cooperation. At the meeting of prime ministers in Bratislava in May 2014, the Polish prime minister expressed his negative stance towards the Hungarian partner, claiming that V4 cooperation is more than a symbolic representation of the common past and future and the threat from Russia cannot be ignored [23,24].

The European Union was a leader in the UNFCCC (United Nations Framework Convention on Climate Change) negotiations performed from 1997 to 2005. The details regarding the implementation of the Kyoto Protocol have been agreed upon by the countries. Individual greenhouse gases (GHG) emission reduction targets for developed countries were negotiated, and their entry into force was realized [25,26]. The EU plays an important role in securing a successful outcome of this process: it played a protagonist role when the United States decided not to ratify the protocol [27].

From the perspective of 2021, the existence of the Visegrad Group turned out to be a mechanism for consulting the positions of the countries in the region before the proceedings of the relevant bodies of the European Union. Within the Visegrad Group, Poland maintains the status quo, simultaneously striving for the development of the Three Seas Initiative and the Bucharest format. The most important issues of the recent period in Poland's EU policy were as follows: the adoption of the EU's multiannual financial framework by the member states as well as the issue of how to introduce carbon neutrality and, above all, the issue of cost sharing of this process between all participants. The importance of these issues in Poland is underlined by the decisions to establish a separate ministry for climate affairs and the separation of European issues from the Ministry of Foreign Affairs. The other V4 countries are natural partners for Poland in the above-mentioned issues. Moreover, there must be consent throughout the EU in both of these issues [28,29].

This manuscript is an attempt to analyse the implementation of obligations under the Paris Agreement by the member states of the Visegrad Group, with particular emphasis on the issue of Poland.

This study aimed to (1) analyze the possibility of fulfilling the Paris Agreement by the Visegrad Group countries (V4); (2) to examine strategies of using by each country of key energy resources; (3) to scrutinize sectors with the largest emission of carbon dioxide in V4 Group, and (4) to show recommendation for European Union and each country of V4 to be on a pathway below 2 °C temperature increase.

The study was based on official reports on CO_2 emissions and obligations resulting from the signed agreements on its reduction.

The multi-threaded research was carried out by analysing not only changes in the structure of emission sources and energy efficiency issues over the years, but also current and future activities aimed at the fulfillment of international agreements in individual V4 countries.

2. Review of Energy Policies of Member Countries of the Visegrad Group

Fossil fuels dominate the Polish energy sector [30]. The country also has the least diversified energy mix in the EU [31,32]. The role of coal in the Polish economy was defended by all governments after 1989. In Poland, the total global energy consumption in

2019 amounted to 4405.8 PJ [33]. This figure slightly differs from the European average. Gross domestic energy consumption per capita in Poland in 2018 amounted to 117.7 GJ (in 2017: 115.9 GJ), with the EU average amounting to 136.0 GJ. An increase in global consumption in 2019 compared to the previous year was observed in the case of crude oil, natural gas, renewable energy and other carriers, and a decrease—in the case of hard coal and lignite. The share of hard coal was 37.0% (in 2018: 40%, in 2017: 40%), lignite 9.1% (in 2018: 10.5%, in 2017: 11%), crude oil 26.3% (in 2018: 25.6% in 2017: 24.6%), natural gas 16.1% (in 2018: 15%, in 2017: 14.8%), renewable energy carriers 9.3% (in 2018: 8.2%, in 2017: 8.1%), and the remaining 2.2% (in 2018: 0.7%, in 2017: 1.5%).

The most important acquired energy carrier in 2019 was hard coal, with a share of 56.2% (57.9% in 2018, 57.9% in 2017). In 2019, over 61.6 million tons of hard coal were mined in Poland (1.8 million tons less than in the previous year). The second largest carrier was lignite with a share of 15.2% (18.1% in 2018, 18.7% in 2017). More than 71% of coal is used to produce heat and electricity [34]. However, Poland depends on the import of this energy carrier, as its domestic production of coal is small [35]. The share of hard coal and lignite in total consumption decreased by 12.8% in 2017 compared to 2000 [36]. The next largest source of energy from TPES (total primary energy supply) is natural gas [37]. In 2019, the share of natural gas in extraction was 5.5% (5.5% in 2018, 5.3% in 2017). One third of natural gas is produced domestically and the remainder is imported. The next largest source of energy is crude oil, with a share of 1.5% (1.6% in 2018, 1.5% in 2017). The remaining renewable energy sources account for 18.3% (16.9% in 2018, 16.6% in 2017). Primary energy obtained in Poland in 2017 amounted to 2723.7 PJ, in 2018 it was 2607.4 PJ, and in 2019 it reached the value of 2528.5 PJ. Poland does not use nuclear energy; however, a scenario for this source of energy in the energy outlook is taken into account [38]. Nuclear energy is a regular theme in public statements. However, this is a challenge due to time-consuming construction, high expenditure and controversy over countries that would provide the required technology. Primary energy consumption in 2019 was 1188 TWh (in 2018, 1170.5 TWh).

The sector of the economy which had the highest share in direct energy consumption was industry (34.6%), while this share was characterized by slight fluctuations in recent years. The second largest sector in terms of consumption was the transport sector, also including private passenger cars—the share of this sector systematically increased and amounted to 28.1% in 2019. In 2019, households used 21.7% of energy, agriculture 4.5%, construction 1.7%, and other recipients—9.4%. The share of coal in electricity production in 2019 was 73.9% (Figure 1). The import of electricity to Poland almost doubled. In 2019, it amounted to 10.6 TWh. Electricity production in Poland was 164 TWh in 2019 (158.5 TWh in 2020).

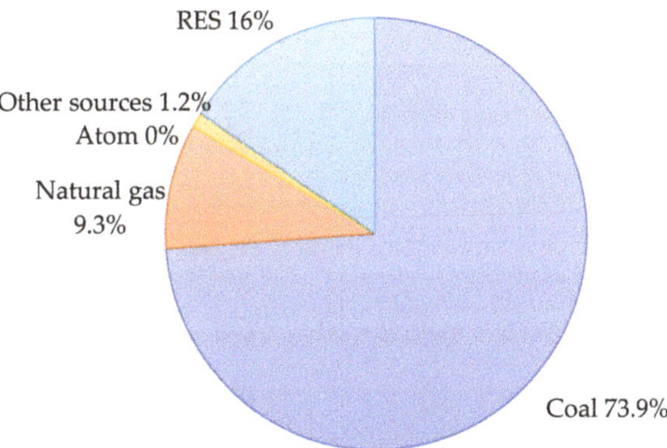

Figure 1. Net electricity generation structure in Poland in 2019.

In the Czech Republic, like in Poland, the dominant source of primary energy in 2018 was coal, which accounted for approx. 36.2% of the total primary energy supply (22.4 Mt in total, of which approx. 5.7 Mt is hard coal, and 16.4 Mtce is brown coal) [39]. The next primary energy carriers were fossil gas (15.8%, 9.7 Mtce) and crude oil (21.6%, 13.3 Mtce). The primary energy mix also includes nuclear energy with a share of 18.1% in 2018 (11.2 Mtce) as well as biofuels and waste which together accounted for 10.2% (6.3 Mtce). The remaining 0.9% of solar, hydro and wind energy provided the remaining 0.6 Mtce. Despite significant dependence on coal energy, the Czech Republic plans to quickly depart from such an energy model [40]. Gradual replacement of fossil fuels (mainly coal) with nuclear energy is being observed, which will favourably affect the GHG emissions. It is assumed in Government projections that the share of nuclear energy will rise from 25% to 33% of TPES by 2040. The Czech Republic is making efforts in the EU forum to qualify nuclear energy as a green energy source. Such a change in qualification would make it possible to obtain funds for the expansion of nuclear energy.

In 2019, electricity was mainly generated from coal—44.2% (51% in 2017) and nuclear energy 34.6% (33% in 2017) (Figure 2) [41,42]. In 2018, the share of coal (lignite and hard coal) decreased to 49%, and nuclear fuel increased to 34% [43]. Additionally, energy production from fossil natural gas accounted for a 6.8% share in 2019 (4.3% in 2018), while 12.7% came from renewable energy sources (11.8% in 2018). In 2018, the Czech Republic generated almost 88 TWh of energy. The Czech Republic was the fourth largest net exporter of electricity in the EU in 2018, after France, Germany and Sweden. Most of its electricity is exported to Austria, Slovakia and Germany [44].

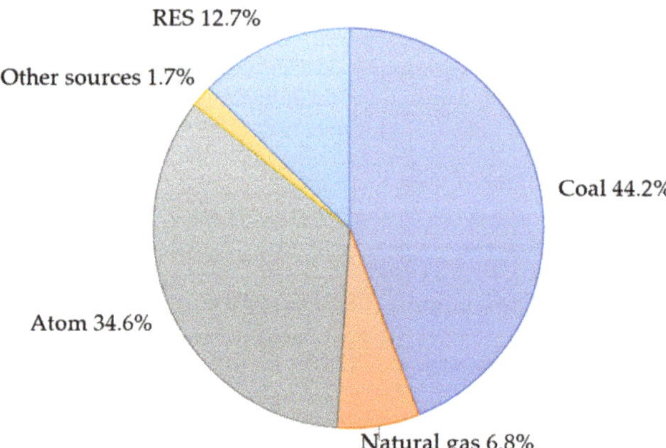

Figure 2. Net electricity generation structure in the Czech Republic in 2019.

In 2019, households (29%), transport (28%) and industry (excluding construction; 27%) had the largest share in final energy consumption. The service sector accounts for 13% of final energy consumption.

The Hungarian energy system differs significantly from the Polish and Czech systems [45,46]. The largest energy carriers are natural gas and crude oil—31.3% and 28.3% of TPES, respectively. Nuclear power is the third largest source (15.6% TPES), and serves for electricity production. In 2018, the total primary energy supply was at the level of 25.2 Mtoe [47]. Coal is not so popular in the Hungarian energy system; it constitutes only 8.5% of the primary energy supply [48]. The most energy-intensive Hungarian sectors are industry and housing, which account for over 60% of total energy consumption—31.6% and 31.5%, respectively. In terms of energy consumption, the third sector is transportation, accounting for 22.3% of TFC. The services sector, including agriculture, accounts for 14.7% of TFC [49–51]. Hungarian energy supplies are dominated by imports from Russia. The country imports about 90% of its crude oil and natural gas, mostly from Russia.

In 2020, 48% of electricity in Hungary came from nuclear power plants (in 2019: 48.2%, in 2018: 49%), 36% of electricity production in Hungary waś based on fossil fuels (25% gas, 11% coal) (Figure 3) [52–54]. For comparison, in 2018, 23% of electricity came from gas and 15% from coal. In 2019, coal was the source of 11.6% of electricity and gas of 25.1%. In 2020, 15% of electricity came from renewable sources (12% in 2018). Electricity production from renewable energy sources is increasing, in line with the EU's Green Electricity Directive. Electricity production in 2019 was 87 TWh. Internationally, the Czech Republic continues to export more electricity than it imports [55].

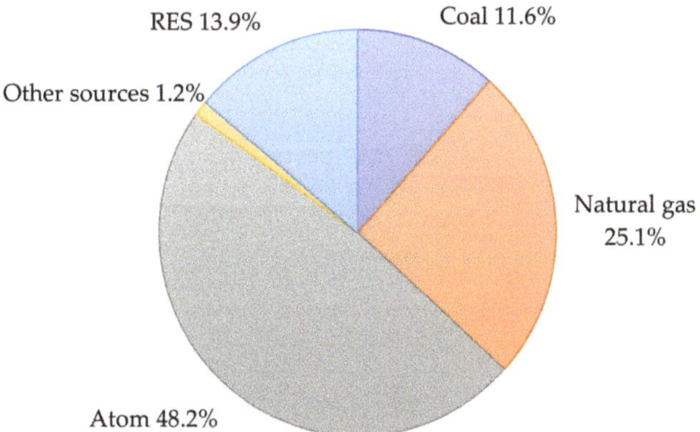

Figure 3. Net electricity generation structure in Hungary in 2019.

In Slovakia, the energy system is dominated by nuclear energy and natural gas, whose shares are 22.7% and 23.5% of TPES, respectively [39,50,51]. In 2016, energy consumption was at the level of 16.5 Mtoe of the total primary energy supply; in 2017, it was 18.0 Mtoe [39]. The energy carriers are mainly crude oil and coal, whose shares is 21.4% and 19.6% of TPES, respectively. In addition, in Slovakia, the highest energy intensity can be observed in industry, with its total energy consumption of about 4 Mtoe. Transportation (about 2.5 Mtoe TFC) and the housing sector (about 2.3 Mtoe) can be distinguished in the second place. Services take the last place in the energy intensity structure, with energy intensity of around 1.5 Mtoe [56,57].

Nuclear power plants are the main contributors to electricity production in Slovakia (Figure 4) [58]. In 2020, 54% of electricity in Slovakia came from nuclear power plants (54% in 2019), 8.5% from coal (8.5% in 2019), 10.2% from natural gas (10.2% in 2019), and 23.3% % from renewable energy sources (23.3% in 2019, Figure 3). The total installed capacity with all energy sources was 7728 MW in 2019 and 7721 in 2017 [59].

Table 1 presents the listing of the energy mix of the Visegrad Group in 2019 [60–62].

Table 1. Sources of electricity in the Visegrad Group states in 2019 r.

Energy Source	Poland	Czech Republic	Hungary	Slovakia
Coal	73.9%	44.2%	11.6%	8.5%
Natural gas	9.3%	6.8%	25.1%	10.2%
Nuclear	0.0%	34.6%	48.2%	54.0%
Other sources	1.2%	1.7%	1.2%	4.0%
RES	15.6%	12.7%	13.9%	23.3%

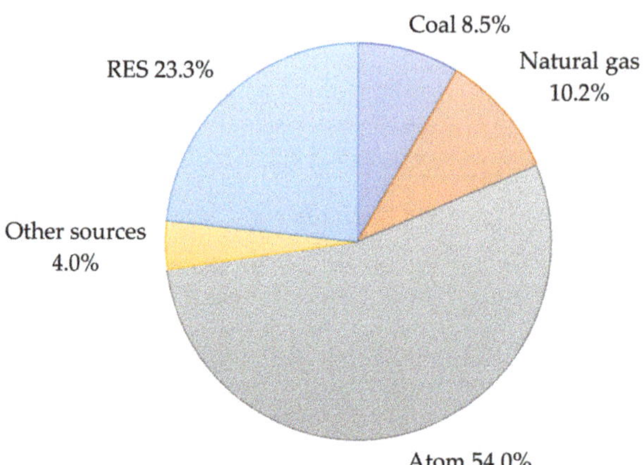

Figure 4. Net electricity generation structure in Slovakia in 2019.

Table 2 presents the listing of the quantity of primary energy consumption in the Visegrad Group states [60–62].

Table 2. Primary energy consumption in the V4 states [in Mtoe].

Country	2011	2012	2013	2014	2015	2016	2017	2018	2019
Poland	100.4	97.4	97.6	94.0	95.0	99.2	99.8	100.7	102.2
Czech Republic	42.9	42.6	41.8	40.9	40.2	39.6	41.3	41.3	40.7
Hungary	23.5	21.9	20.8	20.8	21.9	22.2	23.3	23.5	23.7
Slovakia	16.8	16.1	16.5	15.4	15.5	15.6	16.5	16.2	15.7

Table 3 presents the listing of the quantity of energy generated from renewable sources in the Visegrad Group States [in Mtoe] [60–62].

Table 3. Quantity of energy generated from renewable sources in the V4 states [in Mtoe].

Country	2011	2012	2013	2014	2015	2016	2017	2018	2019
Poland	0.93	1.28	1.26	1.52	1.79	1.78	1.85	1.69	1.99
Czech Republic	0.45	0.51	0.56	0.62	0.66	0.64	0.67	0.67	0.68
Hungary	0.21	0.21	0.22	0.25	0.26	0.26	0.28	0.30	0.37
Slovakia	0.10	0.12	0.13	0.17	0.19	0.20	0.19	0.19	0.16

2.1. Renewable Energy Sources in the Visegrad Group States

In 2018, the EU adopted new rules on renewable energy sources. The RED II Directive sets out a common framework for the promotion of renewable energy and sets a binding EU target of at least 32% share of renewable energy in the EU's gross final energy consumption in 2030. Thus, the EU's climatic and energy goal is to reduce greenhouse gas emissions by at least 40% by 2030, compared to 1990 levels. The European Green Deal already provides for the possibility of increasing the greenhouse gas emission reduction target in the EU not by 40, but by 50–55%, compared to 1990. In line with the guidelines of the new directive, member states were required to define their projected contribution to the achievement of the new EU energy targets. The national energy and climate plans (NECPs) reflect the percentage of renewable energy for the V4 countries such as Poland (23%), Hungary (21%), the Czech Republic (22%) and Slovakia (19.2%) by 2030.

In Poland, the use of renewable energy sources in the energy market has increased significantly since the 1990s [63,64]. The share of energy from renewable sources in the acquisition of primary energy TPES increased in 2011–2019 from 8.2% to 15.6% in total. The structure of energy acquisition from renewable sources for Poland results mainly from the geographical conditions characteristic for Poland and the resources that can be managed [65,66]. The energy obtained from renewable sources in Poland in 2019 came mainly from solid biofuels (65.56%), wind energy (13.72%) and liquid biofuels (10.36%). The total energy value of primary energy obtained from renewable sources in Poland in 2019 was 396,498 TJ [67].

In 2019, Poland produced the most electricity from renewable energy sources in history (over 25 TWh). At the end of 2019, 9.5 GW was installed in RES, of which 1.5 GW in photovoltaic installations. The development of RES in the last two years is mainly the result of investments in prosumer installations. The share of RES in the production of electricity was 15.6%. It is the highest ever (Figure 5).

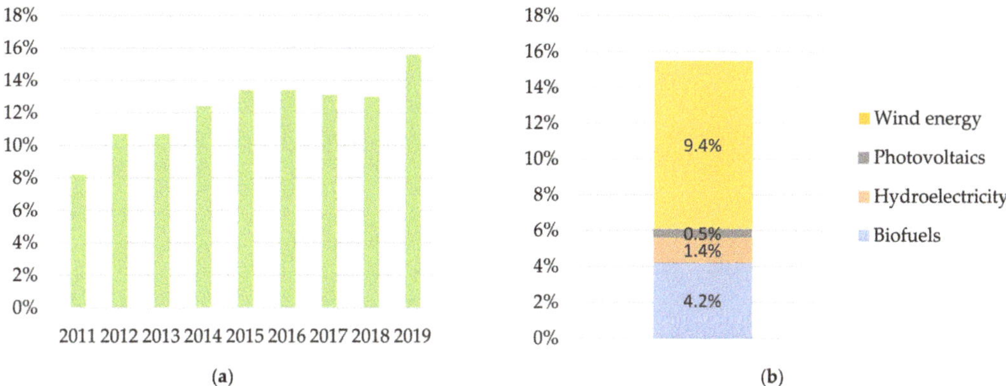

Figure 5. Electricity from RES: (**a**) share of energy from renewable sources in electricity generation in Poland; (**b**) shares of individual RES installations built in Poland in 2019.

In the field of electricity, Poland focuses on the construction of wind farms, but also photovoltaic power plants. The increase in photovoltaics was ensured mainly by reducing the tax for small power plants from 23% to 8% as well as through state subsidy program for these projects.

In January 2020, the Government of the Czech Republic approved the National Energy and Climate Plan for 2021–2030. The document, drawn up by the Ministry of Industry and Trade, introduces certain changes regarding the use of individual fuels in the structure of energy production and consumption in relation to the assumptions set out in the country's energy concept of 2015. The program of the newly adopted strategy assumes an increase in the share of energy from renewable sources in final energy consumption within ten years. The key part of the adopted energy and climate plan is to determine the Czech Republic's contribution to the European climate and energy goals in terms of reducing greenhouse gas emissions, increasing the share of renewable energy sources in energy production and consumption, and increasing energy efficiency. Despite the EC's recommendations to increase the share of renewable energy sources in total energy production and use to a minimum of 23% within a decade, the newly adopted version of the plan from January 2020 assumes that it will reach 22% in 2030.

Renewable energy in the Czech Republic in 2015 reached the level of 3.8 Mtoe, or 9.4% of TPES. The dominant renewable energy sources are biomass and biofuels, which accounted for 8.6% of TPES (3.5 Mtoe). This value increased to 10.5% in 2017. Solar energy, water energy and wind energy took marginal positions in the Czech energy sector with their shares of 0.5%, 0.2% and 0.1% of TPES, respectively. There is no energy production from geothermal sources; however, this energy acquisition source is subject to ongoing research [68–70]. Although the use of renewable energy in the Czech Republic increased from 2% of TPES in 2000 to about 10% in 2017, nuclear power is an essential element of the Czech diversification strategy [68]. Biofuels and biomass account for 10.2% of TPES, while coal constitutes only 8.5% of the primary energy supply. The remaining renewable energy sources amount to only about 1.3% of TPES [47,71].

According to the national climate and energy plan, approved by the Czech Government in January 2020, the share of RES in gross final consumption should increase to 22% by 2030. In 2019, this level was 16%, a large part of which was heating households with biomass. Renewable energy sources accounted for only 12.7% of electricity generation (biofuels 5.8%, hydroelectricity 2/7%, photovoltaics 3.4%, wind energy 0.8%) (Figure 6) [72–75].

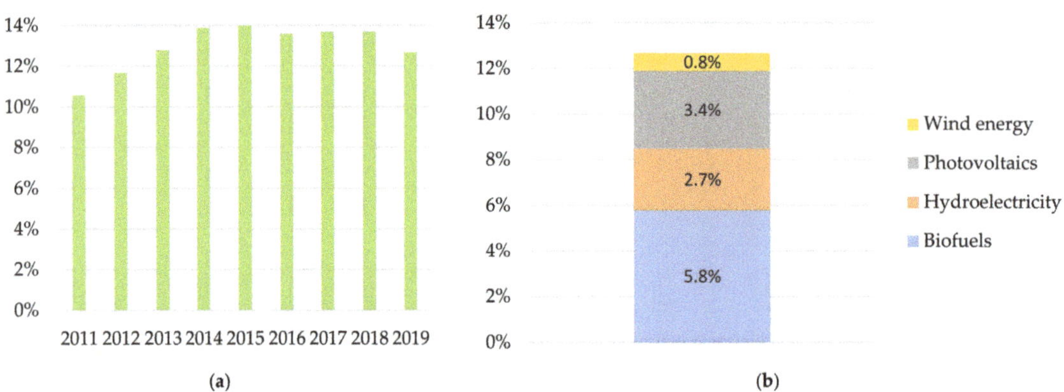

Figure 6. Electricity from RES: (a) share of energy from renewable sources in electricity generation in the Czech Republic; (b) shares of individual RES installations built in the Czech Republic in 2019.

In Hungary, an increase in renewable energy sources in the energy system has been observed over the last ten years. This increase is caused by the use of biomass to produce heat and energy. Biomass is perceived as the source with the highest potential for heat and electricity, while biofuels are the most advantageous alternative to fossil fuel used in transportation [48,49]. In 2018, about 10.1% of the total covered primary energy supply came from renewable energy sources [48,49]. Hydro energy is subsequent with a 1.8% TPES indicator. Other renewable resources, such as geothermal, sun and wind, had negligible shares in the Hungarian energy market (up to 1% of TPES) [47–49]. The amount of electricity produced in Hungary in 2001–2019 decreased from 36.4 TWh in 2001 to 33.9 TWh in 2019. Renewable electricity generation is a growing sector in Hungary. According to 2019 data, the share of renewable energy sources in final electricity consumption was 13.9%, with biofuels (6.1%) being the main type of renewable energy source, followed by photovoltaics 4.9%, wind energy 2.2% and hydropower plants 0.7% (Figure 7). The power of all photovoltaic systems in 2020 was 1170 MW.

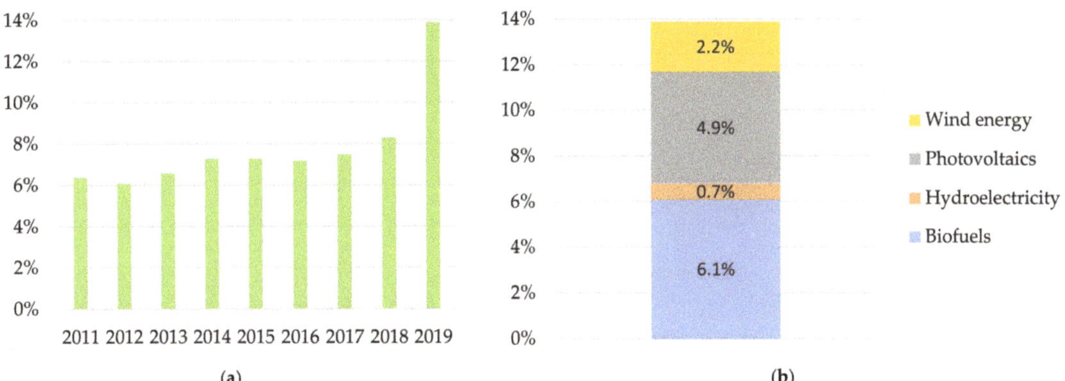

Figure 7. Electricity from RES: (**a**) share of energy from renewable sources in electricity generation in Hungary; (**b**) shares of individual RES installations built in Hungary in 2019.

In Slovakia, about 9.2% of energy carriers' share is attributable to renewable energy sources. Among them, biofuels dominate with a total supply ratio of 8%. The other 1.2% of TPES is water energy [56,57]. The total potential of renewable energy sources in Slovakia in 2019 was around 27,000 GWh per year [76]. Biomass had the greatest technical potential of 11,200 GWh per year, the geothermal potential was around 6300 GWh per year and the geothermal waters had a heating capacity of 280 MW. The technical potential of large hydropower plants was 7600 GWh, and the technical potential of solar energy was estimated at 5200 GWh per year.

In 2019, 6157 GW of electricity was generated from renewable sources, which accounted for 23.3% in the electricity mix (biofuels 3%, hydroelectricity 17.6%, photovoltaics 2.6%, wind energy 0%) (Figure 8) [77–79].

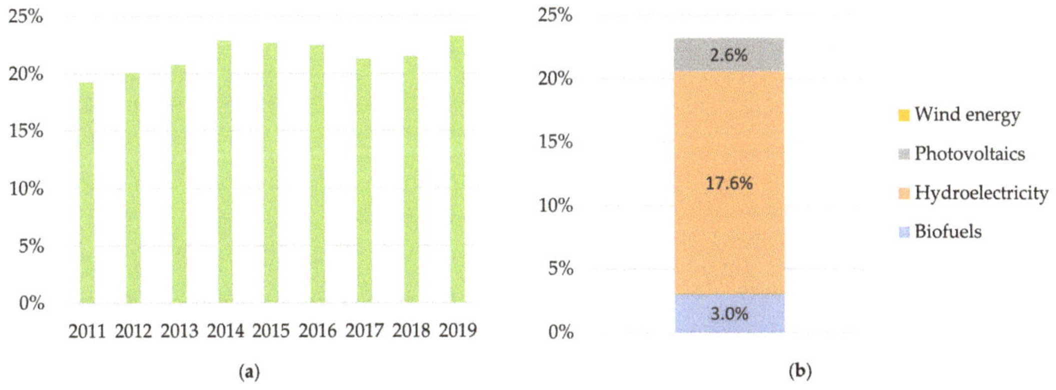

Figure 8. Electricity from RES: (**a**) share of energy from renewable sources in electricity generation in Slovakia; (**b**) shares of individual RES installations built in Slovakia in 2019.

Increasing the share of renewable energy in gross final energy consumption is one of the key objectives of the European Union. This parameter is supposed to gradually bring the EU closer to sustainable development in the field of energy use and has been translated into the goals of individual member states. The goals have been set individually for each EU member state. Table 4 shows the share of energy from renewable sources in the gross final energy consumption (in %) in the countries of the Visegrad Group in the years 2010–2019 [60–62].

Table 4. Share of energy from renewable sources in gross final energy consumption (in %) in the Visegrad Group states in the years 2010–2019.

Country	2011	2012	2013	2014	2015	2016	2017	2018	2019	Goal 2020
Poland	10.3	10.9	11.4	11.5	11.8	11.4	11.1	11.4	12.1	15
Czech Republic	11.0	12.8	13.8	15.1	15.1	14.9	14.8	15.1	16.2	13
Hungary	14.0	15.5	16.2	14.6	14.5	14.3	13.5	12.5	12.6	13
Slovakia	10.3	10.4	10.1	11.7	12.9	12.0	11.4	11.9	16.9	14

Table 5 presents the share of electricity generated from renewable sources in gross final energy consumption in percentage in the Visegrad Group states in 2010–2019 [60–62].

Table 5. Share of electricity produced from renewable sources in gross electricity consumption (in %) in the Visegrad Group states in the years 2010–2019.

Country	2011	2012	2013	2014	2015	2016	2017	2018	2019
Poland	8.1	10.4	10.4	12.5	13.8	13.7	14.2	12.7	15.6
Czech Republic	8.3	9.3	10.8	10.9	11.1	11.4	11.2	11.8	12.7
Hungary	7.5	7.6	9.2	10.7	10.6	10.2	10.6	11.7	13.9
Slovakia	19.3	20.1	20.8	22.9	22.7	24.8	23.9	21.8	23.3

2.2. Energy Efficiency of the Visegrad Group States

All European Union member states are obliged to carry out activities to improve their energy efficiency. One of the priority goals is to increase the share of renewable energy in the produced and consumed energy in Europe.

By the European Parliament and the Council's Directive 2009/28/EC on the promotion of the use of energy from renewable sources, EU member states were obliged to ensure a certain share of energy from renewable sources in their gross final energy consumption in 2020. National targets make up the overall EU target of 20% for the share of renewable energy in gross final energy consumption in the community. For Poland, this target was set at 15%, for the Czech Republic 13%, for Hungary 13% and for Slovakia 14%.

The EU 2030 climate and energy policy framework includes EU-wide assumptions and energy targets for 2021–2030 to reduce greenhouse gas emissions by at least 40% (compared to 1990 levels). The reduction target for Poland, in terms of greenhouse gas emissions in sectors not covered by the ETS system, was set at −7% in 2030 compared to the level of 2005. Moreover, it is planned to increase the share of renewable energy in total gross energy consumption in the EU to at least 32%. As part of the implementation of the EU-wide target for 2030, Poland declares to achieve 21–23% by 2030, the Czech Republic 22%, Hungary 21%, Slovakia 19% of the share of renewable energy in gross final energy consumption (consumption in total in electricity, heating and cooling as well as in transport purposes). In addition, the EU plans to increase energy efficiency by at least 32.5%. The goals of the energy transformation for 2030 in the V4 countries are presented in Table 6.

Table 6. The plan of activity within the framework of energy transformation in the V4 states by 2030.

Country	Share of RES in Final Energy Consumption	Reduction of CO_2 and Greenhouse Gas Emissions in Non-ESTS Sectors	Energy Efficiency
Poland	21–23%	−7%	23%
Czech Republic	22%	−14%	8%
Hungary	21%	−7%	10%
Slovakia	19.2%	−20%	30.3%

For Poland, the national target for improving energy efficiency by 2030 was set at 23% in relation to primary energy consumption according to the PRIMES 2007 forecast, which corresponds to the primary energy consumption of 91.3 Mtoe in 2030.

The main producers of electricity in Poland are conventional utility power plants [80]. They produce about 70 percent of total electricity for distribution and sale in the national energy system. The average age of power plants in Poland is 47 years. Aging power generation units reduce the level of Poland's energy security. Moreover, the problem of the Polish energy generation sector is the relatively low efficiency of coal-based power generation and the accompanying high carbon dioxide emissions. The average efficiency of Polish power plants is lower than that of power plants in the EU. The newest units of the Łagisza II power plant—41%, Pątnów II—41%, Bełchatów II—42%, Opole II—45% have the highest efficiencies, but most of the remaining power plants are characterized by efficiency below 36% (the oldest—even below 30%). With the current efficiency of steam power plants, CO_2 emissions are estimated at approx. 1100–1200 kg CO_2/MWh.

In 2018, the energy production in Poland was 62.4 Mtoe. The final electricity consumption was 166.84 TWh. The total CO_2 emissions were 305.75 Mt.

In Poland, the total final consumption (TFC) of energy has increased over the last decade [81]. In 2017, TFC increased by approximately 20%, compared to 2010 [82]. Poland has implemented the main requirements of the Energy Efficiency Directive (EED) of 2012 (2012/27/EU) by adopting the Energy Efficiency Act (updated on 20 May 2016). The requirements set out in the act include an energy-saving system for energy companies and other measures resulting in 1.5% savings annually, from 2014 to 2020. One of the most important tools for the achievement of the Polish Energy Efficiency Obligations was the 'white certificate system' introduced in 2013. There is also a national action plan with sectoral programs in Poland to support and promote activities to improve energy efficiency [83]. The National Fund for Environmental Protection and Water Management (NFOŚiGW) possesses financial resources to allocate energy efficiency in public and private construction sectors. The fund also provides a national energy advisory system at the regional and local levels [84,85]. The most important documents that define the energy efficiency policy until 2020 include the Polish Energy Policy until 2030 and the National Action Plans on energy efficiency, the development of which was required by Directives 2006/32/EC and 2012/27/EU. Poland's energy policy in the longer term is presented in these strategic documents: Poland's Energy Policy until 2040 (PEP2040) and the National Energy and Climate Plan for 2021–2030. According to the updated Polish Nuclear Energy Program (PPEJ), published in October 2020, the first energy reactor is to be commissioned in 2033 and a further six at two-year intervals, so that the installed capacity in nuclear power plants is between 6 and 9 GW in 2043. The PEP2040 document assumes an increase in the share of RES in all sectors and technologies. In 2030, the share of RES in gross final energy consumption will be at least 23%—not less than 32% in the electricity sector (mainly due to wind and photovoltaic energy), 28% in heating and 14% in transport with a large contribution of electromobility. Offshore wind energy will reach approx. 5.9 GW in 2030 and approx. 11 GW in 2040, in terms of installed capacity. There will be a significant increase in the installed capacity in photovoltaics, around 5–7 GW in 2030, and 10–16 GW in 2040. The share of coal in electricity production will reach 37–56 percent in 2030 and 11–28 percent in 2040, depending on whether the price of emission allowances increases faster or slower.

The Czech Republic is introducing provisions on its energy policy to support and improve the country's energy efficiency in line with the European Union guidelines. The Czech energy policy emphasized increasing overall energy efficiency in all sectors of the economy. The adopted target was the efficiency increase of 20% by 2020 and a further increase of energy efficiency to reduce energy intensity and average energy consumption per capita below the EU Member States average. The Czech Republic set the National Action Plan for Energy Efficiency, which introduced specific quantitative goals for energy savings. The national target for energy efficiency was set at 1060 PJ of final energy consumption.

The obligation to cumulate energy savings has been set at 204.39 PJ of cumulative energy savings by 2020.

The State Program for Supporting Energy Saving and Renewable Energy Use (EFEKT) supports financial energy efficiency, secondary and renewable energy sources [86,87]. From 2004 to 2017, a decrease by about 30% in the total final energy consumption was observed in the Czech Republic. In the most energy-intensive sector—industry—the trend of energy consumption declined between 2004 and 2014 (−16.9%). Within ten years, the structure of demand for individual energy carriers has also changed to the benefit of renewable raw materials (75.1% TFC for biomass and biofuels).

The Czech Government assumes a gradual withdrawal from coal-based energy with a simultaneous increase in the share of nuclear power in the national energy mix. In 2020, the Czech coal commission set the date of abandonment of coal in the energy sector for 2038 (in accordance with the requirements of the Paris Agreement of 2030). The reduction in the importance of coal in the Czech energy mix is to be compensated by further increasing the share of nuclear power in it. This is supposed to be made possible by the construction of the fifth unit at the Dukovany Nuclear Power Plant—the construction of the unit is scheduled for 2029–2036. The Czech Government plans to increase the share of renewable energy sources in the national energy mix to 20.8% in 2030. Such assumptions, however, differ from the European Commission's recommendation for the Czech Republic (23% in 2030). So far, the Czechs have managed to achieve such goals even before the suggested date. In 2013, the Czech Republic reached the 13% share of RES in its energy consumption, as recommended by the European Commission (Table 4). One of the principal goals of the Czech Government is to achieve energy self-sufficiency of the state. The declarations do not translate into reducing dependence on gas supplies from Russia. This is evidenced, e.g., by support for the Nord Stream 2 gas pipeline, which would supply the Czech Republic via the EUGAL gas pipeline running along the Polish–German border. Poland and the Czech Republic are unanimous in their announcements regarding investments in nuclear energy.

In Hungary, in the last decade, a decrease in final energy consumption is observed, despite increases in energy intensity in the housing sector. Residential and commercial sectors combined account for 46% of TFC, but energy consumption has decreased thanks to measures to improve energy efficiency in construction. Energy consumption in industry and transport has increased in recent years. The Hungarian energy strategy and the national action plan for energy efficiency were developed to improve energy efficiency. One of the priorities of Hungarian financial support programs is to reduce energy intensity. The important programs in improving energy efficiency include the green economy financing program and the operational program in environment and energy efficiency [88–91]. Hungary's energy strategy is based on three pillars: becoming independent from Western energy companies, rebuilding and strengthening state-owned companies in the gas and liquid fuel sector and lowering electricity and gas prices, particularly for households. In the field of energy, Hungary works closely with Russia. The Russians also provided credit for 80 percent of the expansion of the nuclear power plant in Paks, which currently produces nearly 40 percent of electricity consumed in Hungary. The Hungarian economy is low-emission compared to the region, which results from the small share of coal in the energy mix (Table 1). The current, binding energy strategy of Hungary with an outlook until 2030 was adopted in 2011. Hungary has determined the increase of RES in its energy mix in 2020–2030 only to be 20%. In addition, the reduction of CO_2 emissions in the power sector is to be largely based on nuclear energy—the fifth and sixth reactors at the Paks power plant are to be operational by 2027. Hungary plans to reduce GHG emissions in its national climate strategy by 52–85% by 2050, depending on available technologies.

In Slovakia, energy consumption has decreased by approx. 14% since 2002 [92] as a result of implementing modern production and less energy-intense technologies. Although industry is the sector with the highest energy consumption, a certain decrease in energy intensity can be seen since 2009. The steady downward trend is visible for the housing sector (approx. 30%). On the other hand, energy consumption increased in the transport sector (by approx. 18%). The Slovak energy policy has several priority areas: increasing efficiency of the combined heat and power plants, reducing transmission and distribution losses (especially electricity, gas and heat), and improving the efficiency of electricity production from hydroelectric plants [93]. In the housing sector, Slovakia focuses on information campaigns and legal regulations regarding energy performance for building components and equipment, including regular inspections of air conditioners and heating devices. Energy efficiency improvement assumption and goal is presented in the Energy Policy of the Slovak Republic [57,86]. The Slovak Government has declared its support for the EU's goal of achieving climate neutrality by 2050 [94]. Slovakia has committed to achieving an even balance of greenhouse gas emissions and absorption by 2050. According to the Government estimates, 82% of electricity produced in Slovakia in the near future will come from emission-free sources, mainly from nuclear power plants (e.g., from the new units in Mochovce) [95,96]. According to the Government plan, subsidies for the production of electricity from coal will end in 2023, and the last Slovak lignite mines will be shut down in 2027.

The planned dates for the termination of coal-based energy production in the V4 countries are presented in Table 7.

Table 7. The plan of activity within the framework of energy transformation in the V4 states by 2030.

Goal	Poland	Czech Republic	Hungary	Slovakia
Termination of energy production from coal	2030/2040	2038	2049	2023

2.3. The Structure of CO_2 Emissions and Climate Policy of the Visegrad Group Countries

While the EU average of total annual CO_2 emissions has fallen by a fifth since 1990, the pace of decarburisation between countries has varied. Over the past 30 years, Hungary has managed to reduce its emissions by 32%, the Czech Republic by 35%, and Poland by 13%.

Poland is one of the largest gas emitters in the EU. Poland ranks third in Europe in terms of CO_2 emissions, and fifth in terms of all greenhouse gas emissions. By 2018, Poland had managed to reduce greenhouse gas (GHG) emissions by 13 percent compared to 1990 (the EU plan is 40%). According to Global Carbon Budget estimates from 2019, Poland ranks 20th in the world among the largest CO_2 emitters—the highest among all the member states of the Visegrad Group. Poland's CO_2 emissions are estimated at 323 megatons of CO_2. The Czech Republic came 43rd with 101 Mt CO_2, then Slovakia which ranked 75th (33 $MtCO_2$) and Hungary at the 57th place (49 $MtCO_2$). The largest source of emissions is electricity which is responsible for a quarter of emissions in Poland, which is due to the large share of coal in energy production. In 2018, imports to Poland amounted to 19.3 million tons of coal, and a year later 14.9 million. The vast majority of the Polish energy sector (about 70%) is still based on coal. The biggest polluter in Europe is the Polish power plant Bełchatów, which emitted 38.2 megatons of CO_2 in 2018 alone. Transport plays an increasingly infamous role, accounting for about 15% of emissions. Industry and processing each account for 8 percent in the emission structure. The actions taken in the country show that only the implementation of four flagship projects will possibly reduce greenhouse gas emissions by 42 percent. This means that the missing gap in relation to the EU target is between 2 and 9 percent. The aforementioned flagship projects cover areas that are still underway, due to the ongoing economic processes or e.g., the ongoing fight against smog in Poland. They include: a change in the electricity mix, transformation in heating, innovation in the industrial sector and electrification of transport. Poland already has emission reduction plans in most of these areas.

Based on the annual reports on CO_2 emissions for 2020 from ETS participants, the total amount of CO_2 emissions in Poland covered by the EU ETS in 2020 amounted to 172.15 million tons of CO_2, including the aviation sector.

The total annual greenhouse gas emissions in the Czech Republic in 2018 were 129.39 million tons (in 2019, they were 131 million tons of CO_2). The Czech emissions represent 3.5% of the total EU emissions and have decreased by almost 13% since 2005. It should be noted that the average reduction of emissions across the EU was 19% during the same period. Emissions in the energy sector amounted to 51.07 million tons of CO_2 (39.5% of total emissions), in the transport sector 20.3 million tons of CO_2 (15.7% of total emissions), emissions from industrial processes 16.26 million tons of CO_2 (12.6% of total emissions), emissions from fossil fuel combustion for industry 9.96 million tons of CO_2 (7.7% of total emissions), combustion in households, institutions and agriculture generated 13.15 million tons of CO_2 (10.2% of total emissions), emissions from waste management 5.7 million tons of CO_2 annually (4.4% of total emissions), agricultural emissions 8.61 million tons of CO_2 (6.7% of total emissions).

The climate protection policy of the Czech Republic was adopted by the Czech Government in March 2017 and replaced the previous national program to mitigate the effects of climate change in the Czech Republic. The main goals presented in the document are as follows: the reduction of greenhouse gas emissions by 32 Mt CO_2eq by 2020, compared to 2005 and the reduction of greenhouse gas emissions by 44 Mt CO2eq by 2030, compared to 2005. In terms of energy security, the target is to keep energy import dependency below 65% by 2030 and 70% by 2040 (nuclear fuel is considered an imported resource).

The Hungarian Parliament passed the Climate Protection Act in 2020. According to this document, Hungary will reduce its carbon dioxide emissions by at least 40 percent by 2030, compared to 1990 levels, and renewable energy will have accounted for at least 21 percent of consumed energy by then. In addition, Hungary is to be entirely climate neutral by 2050. The Hungarian Government has decided to pin its hopes on nuclear power (supplemented by increased photovoltaic capacity) to meet emission reduction targets and increase the rate of emission-free electricity production from the current 60 percent to 90 percent by 2030. Currently, it is to be investigated how low-carbon nuclear energy can be used to produce clean hydrogen. According to data from the EU Statistical Office, Hungarian emissions dropped by 0.8 percent in 2018 and amounted to about 1.4% of total EU CO_2 emissions.

The emission of carbon dioxide from Hungary was 58 Mt CO_2eq in 2014 [88] and 45.8 Mtoe CO_2eq in 2017 [39]. Hungary reduced greenhouse gas emissions by 32% between 1990 and 2017, however, these emissions have been increasing. In 2020, Hungary generated 45.5 metric tons of carbon dioxide emissions. That was a drop by almost 4 percent, compared to 1990 levels. Greenhouse gas emissions have decreased significantly due to the transformation in the economic and energy sectors in reduced consumption of fossil fuels. It is energy industry (28%)—mainly heating, based on natural gas in 80%—and transport (26%) that are the most responsible for those emissions. Due to the high share of emissions from the transport sector, the National Transport Strategy was developed, focusing its activities on broadly understood mobility. It sets targets for increasing the number of transport means with low GHG emissions and obliges to increase the share of biofuels in the sector. It also indicates the need to share electricity and hydrogen sources in overall energy consumption. Hungary also uses ETS emissions trading systems, which are used to finance emission reduction support schemes [48,49].

Slovakia generated 35.9 Mt CO_2 in 2019 (was 38.1 Mt CO_2 in 2018). The sector with the largest CO_2 emission was energy production, responsible for 50% of all emissions. The industry sector was responsible for 23%. A total 16% of emitted CO_2 originated from transport, 7% of emissions resulted from agriculture and 4% from waste. By 2023, Slovak coal power plants and all lignite mines will have been shut down. Only one will be allowed to mine lignite for non-energy purposes.

The basic documents for the development of climate change mitigation policies are "Low-Emission Development Strategy of the Slovak Republic until 2030 with a Prospect to 2050" and the "Integrated National Energy Plan of Slovakia until 2030". The low-emission development strategy of the Slovak Republic until 2030 with a prospect to 2050 was approved by the Government of the Slovak Republic in March 2020. The main objective of the document was to set the direction for achieving climate neutrality in the Slovak Republic by 2050. The strategy includes national emission reduction targets by 2030, based on European targets. These goals were specified in the integrated national energy plan of Slovakia until 2030: the share of RES 19.2% (EU targets 32%) and energy efficiency at 30.3% (EU targets 32.5%).

3. Summary

In 2020, the share of European electricity production from renewable sources increased to 38% (compared to 34.6% in 2019). Conversely, electricity generation from fossil fuels has fallen to 37%. It is undoubtedly an important moment in the European transformation to clean energy. Still, the pace of the energy transition is too slow to achieve a 55% reduction in greenhouse gas emissions by 2030 and climate neutrality by 2050. Unfortunately, according to new data from the Ministry of Energy, Poland will not achieve the expected share of renewable energy in 2020. The share of renewable energy sources in energy consumption in 2018 was 11.4%, and in 2019 it was 12.1%. The national target for 2020 was 15%. The share of RES in Poland in 2020 was 13.8% instead of 15%. In 2019, the share of the Czech Republic in RES was 16.2% (the national target for 2020 was 13%), Hungary had 12.6% (the target was 13%) and Slovakia achieved 16.9% (the target was 14%).

This means that Poland only keeps emissions at the bottom of the target. Of the 28 EU countries, only 12 have reached the national target for the distribution of renewable energy sources by the end of the energy supply set for 2020 [89].

The policies and actions currently described in the National Forecast are insufficient to achieve the steps needed to reach the EU's target of at least 40% reduction by 2030. Depending on the existing actions in the country [89], this problem will be mitigated by 26% from 2020, and 30% until 2030. To comply with the Paris Agreement, the EU will have to exceed the 32% renewable energy rate by 2030. This would increase the RES energy surplus to 55%, as the share of renewable energy is up to 59%. By 2030—75%, by 2050, the total decarbonization will be achieved [93]. Renewable energy accounts for 17% of total energy consumption in the EU-28, with a target of 20%.

In November 2018, the European Commission stated the European climate strategy for 2050 titled "A Clean Planet for All." These EU Member States have adopted or offered long-term abandonment strategies. All EU Member States should prepare and report on a long-term strategy with a vision of at least 30 years [97].

The EU ambition concerning mitigation has been relatively limited, mainly due to the group of Central and Eastern European EU member states led by Poland, and the Visegrad Group [15]. The EU's GHG emission reduction target for 2030 is 40%. It was constituted based on INDC and was agreed upon in October 2014; however, the Visegrad Group was against more ambitious targets. It is worth saying that it is not ambitious enough to keep the world on a pathway below a 2 °C temperature increase. The EU is trying to force other developed countries to make ambitious commitments on climate change mitigation after 2020 [98,99].

The development of the European Union's climate and energy policy in 2018 means that the EU can gain the EU's global leadership position in negotiations on climate action. The currently implemented climate action level is not yet compatible with the Paris Agreement's 1.5 °C limits. The European Parliament has called for increasing the EU's INDC emissions 2030 reduction goal to 55% below 1990 levels. Achieving this would need to build on the reform of the EU ETS, the adoption of the Energy Performance of Buildings Directive (EPBD), and the political agreement on increasing renewable energy and energy efficiency targets for 2030 [93].

The recent disagreements between Poland and Hungary by the Visegrad Group on the EU's energy and climate agenda are a good sign (on Russia). Divided, the association can hardly thwart major environmental policies. In particular, Poland will be more difficult to defend against as a coal miner that does not meet EU energy and climate targets. Suppose the EU and its negotiating members succeed in exploiting this dangerous alliance. In this case, this could bring the association closer to the winter solstice under the Paris Agreement [93].

In summary, three of the Visegrad Group (except Poland) countries meet the commitments made in reducing carbon dioxide emissions. The consequence of failing to meet these commitments is rising emissions trading costs, energy costs and production costs (closure of production facilities or temporary shutdown of production, for example, Huta Kraków) [100].

4. Conclusions

In 2021, the Visegrad Group celebrates its 30th anniversary. The group has included the Czech Republic, Poland, Slovakia and Hungary since 1 January 1993, following the break-up of Czechoslovakia. Its members are linked by neighbourhood and similar geopolitical conditions. The keynote of establishing the group was cooperation on strengthening the processes of democratic transformations of state structures and building a free market economy, so that in the longer term it would be possible to achieve the goal of European and Atlantic integration (all V4 countries have been members of the European Union and NATO since 2004 and 1997, respectively). Currently, the Visegrad Group is a forum for exchanging experiences and working out common positions on issues important for the future of the region and the EU. In 2015, almost 200 countries decided to conclude a global agreement (the Paris Agreement) to combat global warming. This agreement was also signed by the countries of the Visegrad Group. For Poland, whose energy sector is based primarily on coal, any provisions regarding the reduction of CO_2 emissions are a big challenge.

The manuscript presents the situation of the energy sector of the Visegrad Group countries in recent years. The climate and energy policy in recent years was analysed, as well as the plans of the V4 countries in the context of the European Union's climate goals for 2020, 2030 and 2050. The level of energy efficiency, the structure of the shares of individual energy carriers, the share of renewable energy sources and the emission of carbon dioxide have been analyzed in detail and in many ways.

Slovakia, the Czech Republic and Hungary are countries that have already achieved the level of emission reductions required by the convention (compared to the reference year) and can contribute to the achievement of the EU target. Only Poland would keep its emissions below the target. It should be emphasized that Poland has ratified four international agreements on climate protection: the Climate Convention of 1994, and the Kyoto Protocol of 2002. In 2016, it also ratified the Paris Agreement, and in 2018, an amendment to the Kyoto Protocol, which expires after 2020. In the last two, Poland undertook to meet the reduction targets together with the EU.

The following list contains a collection of the most important names used in the manuscript, along with the corresponding symbols (Table 8).

Table 8. List of abbreviations and nomenclature.

Parameter	Description
COP	Conference of the Parties
CO_2	Carbon dioxide
EED	Energy Efficiency Directive
ETS	Emission Trading Scheme
ESE	Energy and Environment Safety Strategy
GEFS	Green Economy Financing Program
GHG	Greenhouse gases
HFC	Hydrofluorocarbon
INDC	Intended Nationally Determined Contributions
IPCC	Intergovernmental Panel on Climate Change
Mtoe	Million tons of oil equivalent

Table 8. *Cont.*

Parameter	Description
Mt CO_2 eq	Metric tons of carbon dioxide equivalent
NF_3	Nitrogen trifluoride
NFOŚiGW	The National Fund for Environmental Protection and Water Management
NTS	National Transport Strategy
PFC	Perfluorocarbon
RES	Renewable energy sources
SEnvP	Slovak National Environmental Policy
SF_6	Sulfur hexafluoride
TPES	Total primary energy supply
TFC	Total final consumption
UNFCCC	United Nations Framework Convention on Climate Change
V4	Visegrad Group
WCS	White certificate system

Author Contributions: K.T., I.W., O.O., A.K., A.B., M.K. conceptualization, K.T., I.W., A.K., A.B., M.K. methodology, K.T., I.W., A.K., A.B., O.O. writing—original draft preparation, K.T., M.K. review and editing. All authors have read and agreed to the published version of the manuscript.

Funding: This research received no external funding.

Institutional Review Board Statement: Not applicable.

Informed Consent Statement: Not applicable.

Data Availability Statement: Not applicable.

Acknowledgments: M.K. would like to acknowledge the financial support provided her for this study by the project financed by the Ministry of Science and Higher Education in Poland under the program "Regional Initiative of Excellence" 2019–2022 Project Number 015/RID/2018/19 total funding amount 10 721 040,00 PLN.

Conflicts of Interest: The authors declare no conflict of interest.

References

1. Paris Agreement. Available online: https://unfccc.int/process#:a0659cbd-3b30-4c05-a4f9-268f16e5dd6b (accessed on 24 May 2021).
2. The Paris Agreement on Climate Change. Available online: https://undocs.org/FCCC/CP/2015/L.9/Rev.1 (accessed on 24 May 2021).
3. Rajgor, G. Greater acceleration of renewables required to meet COP21 goal. *Renew. Energy Focus* **2016**, *17*, 175–177. [CrossRef]
4. Balibar, S. Energy transitions after COP21 and 22. *Comptes Rendus Phys.* **2017**, *18*, 479–487. [CrossRef]
5. Ari, I.; Sari, R. Differentiation of developed and developing countries for the Paris Agreement. *Energy Strategy Rev.* **2017**, *18*, 175–182. [CrossRef]
6. Morgan, E.A.; Nalau, J.; Mackey, B. Assessing the alignment of national-level adaptation plans to the Paris Agreement. *Environ. Sci. Policy* **2019**, *93*, 208–220. [CrossRef]
7. Livingston, J.E.; Lövbrand, E.; Olsson, J.A. From climates multiple to climate singular: Maintaining policy-relevance in the IPCC synthesis report. *Environ. Sci. Policy* **2018**, *90*, 83–90. [CrossRef]
8. Devès, M.H.; Lang, M.; Bourrelier, P.H.; Valérian, F. Why the IPCC should evolve in response to the UNFCCC bottom-up strategy adopted in Paris? An opinion from the French Association for Disaster Risk Reduction. *Environ. Sci. Policy* **2017**, *78*, 142–148. [CrossRef]
9. Liyanage, S.; Dia, H.; Abduljabbar, R.; Bagloee, S.A. Flexible Mobility On-Demand: An Environmental Scan. *Sustainability* **2019**, *11*, 1262. [CrossRef]
10. Dröge, S. The Paris Agreement 2015. Turning Point for the International Climate Regime. Available online: https://www.ssoar.info/ssoar/handle/document/46462 (accessed on 24 May 2021).
11. Nazarko, J.; Czerewacz-Filipowicz, K.; Kuźmicz, A.K. Comparative analysis of the Eastern European countries as participants of the new silk road. *J. Bus. Econ. Manag.* **2017**, *18*, 1212–1227. [CrossRef]

12. International Cooperation under Article 6 of the Paris Agreement: Reflections before SB 44. Available online: http://www.greengrowthknowledge.org/resource/international-cooperation-under-article-6-paris-agreement-reflections-sb-44 (accessed on 24 May 2021).
13. Clemencon, R. The Two Sides of the Paris Climate Agreement: Dismal Failure or Historic Breakthrough? *J. Environ. Dev.* **2016**, *25*, 3–24. [CrossRef]
14. European Commission the Road from Paris: Assessing the Implications of the Paris Agreement and Accompanying the Proposal for a Council Decision on the Signing, on Behalf of the European Union, of the Paris Agreement adopted under the United Nations Framework Convention on Cl. Available online: https://eur-lex.europa.eu/legal-content/EN/ALL/?uri=COM:2016:0110:FIN (accessed on 24 May 2021).
15. Bodle, R.; Donat, L.; Duwe, M. The Paris Agreement: Analysis, Assessment and Outlook. *Carbon Clim. Law Rev.* **2016**, *10*, 5–22.
16. Olgun, C. The role of the Eastern member states in the European Union's energy and climate policy. *Inst. Int. Political Econ. Berl.* **2017**, *89*, 1–46.
17. Sacio-Szymańska, A.; Kononiuk, A.; Tommei, S. Mobilizing Corporate Foresight Potential among V4 Countries—Assumptions, Rationales and Methodology. *Procedia Eng.* **2017**, *182*, 635–642. [CrossRef]
18. Molendowski, E. The Visegrad Group Countries—Changes in Intra-industry Competitiveness of their Economies During the World Financial and Economic Crisis. *Procedia Soc. Behav. Sci.* **2014**, *110*, 1006–1013. [CrossRef]
19. Gikas, G. The aspects of the cooperation within the Visegrad Group. *Appl. Res. Rev.* **2016**, *3*, 219–229.
20. Dorożyński, T.; Kuna-Marszałek, A. Investments Attractiveness: The Case of the Visegrad Group Countries. *Comp. Econ. Res.* **2016**, *19*, 119–140. [CrossRef]
21. Fawn, R. Visegrad: Fit for purpose? *Communist Post-Communist Stud.* **2013**, *46*, 339–349.
22. Minarik, M. Energy Cooperation in Central Europe: Interconnecting the Visegrad Region. Available online: https://www.energycharter.org/fileadmin/DocumentsMedia/Occasional/Visegrad_Cooperation.pdf (accessed on 24 May 2021).
23. Racz, A. The Visegrad Cooperation: Central Europe divided over Russia. *Eur. Form.* **2014**, *4*, 61–76. [CrossRef]
24. Marusiak, J. Russia and the Visegrad Group—More than a foreign policy issue. *Int. Issues Slovak Foreign Policy Aff.* **2015**, *14*, 28–46.
25. Wang, C.H.; Ko, M.H.; Chen, W.J. Effects of Kyoto Protocol on CO_2 Emissions: A Five-Country Rolling Regression Analysis. *Sustainability* **2019**, *11*, 1–20.
26. Kuriyama, A.; Abe, N. Ex-post assessment of the Kyoto Protocol—Quantification of CO_2 mitigation impact in both Annex B and non-Annex B countries. *Appl. Energy* **2018**, *220*, 286–295. [CrossRef]
27. Groen, L. On the road to Paris: How can the EU avoid failure at the UN Climate Change Conference (COP21)? *IAI Work. Pap.* **2015**, *33*, 1–14.
28. Dzikuć, M.; Wyrobek, J.; Popławski, Ł. Economic Determinants of Low-Carbon Development in the Visegrad Group Countries. *Energies* **2021**, *14*, 3823. [CrossRef]
29. Kijewska, A.; Bluszcz, A. Carbon footprint levels analysis for the world and for the EU countries. *Syst. Wspomagania W Inżynierii Prod.* **2017**, *6*, 169–177.
30. Manowska, A.; Tobór-Osadnik, K.; Wyganowska, M. Economic and social aspects of restructuring Polish coal mining: Focusing on Poland and the EU. *Resour. Policy* **2017**, *52*, 192–200. [CrossRef]
31. Fortuński, B. The impact of the Poland foreign trade on its real CO_2 emissions. *Econ. Environ. Stud.* **2017**, *17*, 1161–1174. [CrossRef]
32. Skjærseth, J.B. Implementing EU climate and energy policies in Poland: Policy feedback and reform. *Env. Polit.* **2018**, *27*, 498–518. [CrossRef]
33. Fuel and Energy Economy in 2018 and 2019. Available online: https://stat.gov.pl/obszary-tematyczne/srodowisko-energia/energia/gospodarka-paliwowo-energetyczna-w-latach-2018-i-2019,4,15.html (accessed on 25 July 2021).
34. Nieć, M.; Sermet, E.; Chećko, J.; Górecki, J. Evaluation of coal resources for underground gasification in Poland. Selection of possible UCG sites. *Fuel* **2017**, *208*, 193–202. [CrossRef]
35. PGI-NRI. Oil and Gas in Poland. New Opportunities 2015. Available online: https://infolupki.pgi.gov.pl/sites/default/files/czytelnia_pliki/folder_na_bruksele.pdf (accessed on 24 May 2021).
36. Główny Urząd Statystyczny. Green Economy Indicators in Poland 2019. Available online: https://stat.gov.pl/en/topics/environment-energy/environment/green-economy-indicators-in-poland-2019,3,3.html (accessed on 24 May 2021).
37. Rybak, A.; Manowska, A. The future of crude oil and hard coal in the aspect of Poland's energy security. *Energy Policy J.* **2018**, *21*, 141–154. [CrossRef]
38. Conclusions from Prognostic Analyzes for the Energy Sector—Attachment 1 to Poland's Energy Policy Until 2040 (PEP2040)—Project of Ministry of Energy. Available online: https://www.gov.pl/documents/33372/436746/Wnioski_z_analiz_do_PEP2040_2018-11-23.pdf/1481a6a9-b87f-a545-4ad8-e1ab467175cf (accessed on 24 May 2021).
39. International Energy Agency. Key World Energy Statistics. 2019. Available online: https://www.iea.org/reports/world-energy-statistics-2019 (accessed on 24 May 2021).
40. Energy Policies of IEA Countries 2016 Czech Republic Review. Available online: https://webstore.iea.org/energy-policies-of-iea-countries-czech-republic-2016-review (accessed on 25 July 2021).
41. Národní Energetický Mix. Available online: https://www.ote-cr.cz/cs/statistika/narodni-energeticky-mix (accessed on 25 July 2021).

42. Statistical Yearbook of the Czech Republic. Available online: https://www.czso.cz/csu/czso/10n1-05-2005-energy__methodology_b (accessed on 25 July 2021).
43. PRK Partners Alternative Energy & Power 2019—Czech Republic. Available online: https://practiceguides.chambers.com/practice-guides/alternative-energy-power-2019/czech-republic (accessed on 24 May 2021).
44. Aktuální Stav České Energetiky. Available online: http://www.edotace.cz/clanky/energetika-v-cr-quo-vadis (accessed on 25 July 2021).
45. Hartung, K.; Kiss, T. Time for Change! Decentralized Wind Energy System on the Hungarian Market. *Energy Procedia* **2014**, *52*, 38–47. [CrossRef]
46. Sáfián, F. Modelling the Hungarian energy system—The first step towards sustainable energy planning. *Energy* **2014**, *69*, 58–66. [CrossRef]
47. Eurostat. Energy, Transport and Environment Indicators, 2018 ed. 2018. Available online: https://ec.europa.eu/eurostat/web/products-statistical-books/-/KS-DK-18-001 (accessed on 24 May 2021).
48. Energy Policies of IEA Countries: Hungary 2017 Review. Available online: https://webstore.iea.org/energy-policies-of-iea-countries-hungary-2017-review (accessed on 24 May 2021).
49. IEA Hungary. Available online: https://www.iea.org/countries/Hungary/ (accessed on 24 May 2021).
50. Mišík, M. On the way towards the Energy Union: Position of Austria, the Czech Republic and Slovakia towards external energy security integration. *Energy* **2016**, *111*, 68–81. [CrossRef]
51. Martins, F.; Felgueiras, C.; Smitkova, M.; Caetano, N. Analysis of Fossil Fuel Energy Consumption and Environmental Impacts in European Countries. *Energies* **2019**, *12*, 1–11.
52. Analysis of Climate-Energy Policies & Implementation in Hungary. Available online: http://eko-unia.org.pl/wp-content/uploads/2018/06/mini-report-1_Hungary.pdf (accessed on 25 July 2021).
53. Kumar, B.; Szepesi, G.; Čonka, Z.; Kolcun, M.; Péter, Z.; Berényi, L.; Szamosi, Z. Trendline Assessment of Solar Energy Potential in Hungary and Current Scenario of Renewable Energy in the Visegrád Countries for Future Sustainability. *Sustainability* **2021**, *13*, 5462. [CrossRef]
54. Hungarian Central Statistical Office. Hungary. 2019. Available online: https://www.ksh.hu/docs/hun/xftp/idoszaki/mo/hungary2019.pdf (accessed on 25 July 2021).
55. Why Hungary's Power Market Is a Hotspot? Available online: https://internationalfinance.com/why-hungarys-power-market-is-a-hotspot/ (accessed on 25 July 2021).
56. Energy Policies of IEA Countries: Slovak Republic 2018 Review. Available online: https://webstore.iea.org/energy-policies-of-iea-countries-slovak-republic-2018-review (accessed on 24 May 2021).
57. Nové Znenie. Available online: http://www.rokovania.sk/Rokovanie.aspx/BodRokovaniaDetail?idMaterial=23993 (accessed on 24 May 2021).
58. MHSR Návrh Energetickej Politiky Slovenskej Republiky. Available online: http://www.rokovania.sk/Rokovanie.aspx/BodRokovaniaDetail?idMaterial=23993 (accessed on 24 May 2021).
59. Filčák, R. Slovakia and the landscape of climate change policies: Drivers, barriers and stakeholder analysis. *Progn. Práce* **2016**, *8*, 5–22.
60. Slovakia—Electricity Generation. Available online: https://countryeconomy.com/energy-and-environmen/electricity-generation/slovakia (accessed on 25 July 2021).
61. EU Power Sector in 2020. Available online: https://ember-climate.org/project/eu-power-sector-2020/ (accessed on 25 July 2021).
62. Share of Renewable Energy in Gross Final Energy Consumption. Available online: https://ec.europa.eu/eurostat/databrowser/view/t2020_31/default/table?lang=en (accessed on 25 July 2021).
63. Our World in Data. Available online: https://ourworldindata.org/grapher/share-electricity-renewables?tab=table&time=earliest..2011 (accessed on 25 July 2021).
64. Mączyńska, J.; Krzywonos, M.; Kupczyk, A.; Tucki, K.; Sikora, M.; Pińkowska, H.; Bączyk, A.; Wielewska, I. Production and use of biofuels for transport in Poland and Brazil—The case of bioethanol. *Fuel* **2019**, *241*, 989–996. [CrossRef]
65. Kupczyk, A.; Mączyńska, J.; Redlarski, G.; Tucki, K.; Bączyk, A.; Rutkowski, D. Selected Aspects of Biofuels Market and the Electromobility Development in Poland: Current Trends and Forecasting Changes. *Appl. Sci.* **2019**, *9*, 254. [CrossRef]
66. Paska, J.; Surma, T. Electricity generation from renewable energy sources in Poland. *Renew. Energy* **2014**, *71*, 286–294. [CrossRef]
67. Gawlik, L.; Szurlej, A.; Wyrwa, A. The impact of the long-term EU target for renewables on the structure of electricity production in Poland. *Energy* **2015**, *92*, 172–178. [CrossRef]
68. Energy from Renewable Sources in 2019. Available online: https://stat.gov.pl/obszary-tematyczne/srodowisko-energia/energia/energia-ze-zrodel-odnawialnych-w-2019-roku,10,3.html (accessed on 25 July 2021).
69. Czech Republic needs more nuclear units, report shows. Available online: https://www.world-nuclear-news.org/Articles/Czech-Republic-needs-more-nuclear-units,-report-sh (accessed on 6 August 2021).
70. In focus: The Czech Republic's energy market. Available online: https://ceenergynews.com/finance/in-focus-the-czech-republics-energy-market/ (accessed on 6 August 2021).
71. Jursová, S.; Burchart-Korol, D.; Pustějovská, P.; Korol, J.; Blaut, A. Greenhouse Gas Emission Assessment from Electricity Production in the Czech Republic. *Environments* **2018**, *5*, 17. [CrossRef]

72. OECD. Fossil Fuels Support Country Note. Czech Republic. Available online: https://www.oecd.org/fossil-fuels/data/ (accessed on 24 May 2021).
73. Czech Statistical Office. Fuel and Energy Consumption—2019. Available online: https://www.czso.cz/csu/czso/fuel-and-energy-consumption-2019 (accessed on 25 July 2021).
74. Odyssee-Mure. Czechia Profile. Available online: https://www.odyssee-mure.eu/publications/efficiency-trends-policies-profiles/czechia.html (accessed on 25 July 2021).
75. Czech Republic: Increasing the Share of Renewable Energy in the Energy and Climate Plan. Available online: https://ies.lublin.pl/komentarze/czechy-wzrost-udzialu-energii-ze-zrodel-odnawialnych-w-planie-energetyczno-klimatycznym/ (accessed on 25 July 2021).
76. Share of Energy from Renewable Sources in Electricity Generation in Czechia from 2007 to 2018. Available online: https://www.statista.com/statistics/419423/czech-republic-share-of-electricity-from-renewable-sources/ (accessed on 25 July 2021).
77. Lieskovský, M.; Trenčiansky, M.; Majlingová, A.; Jankovský, J. Energy Resources, Load Coverage of the Electricity System and Environmental Consequences of the Energy Sources Operation in the Slovak Republic—An Overview. *Energies* **2019**, *12*, 1701. [CrossRef]
78. Distribution of Electricity Generation in Slovakia in 2020, by Source. Available online: https://www.statista.com/statistics/1236359/slovakia-distribution-of-electricity-production-by-source/ (accessed on 25 July 2021).
79. Energy Resource Guide—Slovakia—Renewable Energy. Available online: https://www.trade.gov/energy-resource-guide-slovakia-renewable-energy (accessed on 25 July 2021).
80. Slovakia Renewable Energy. Available online: https://www.trade.gov/market-intelligence/slovakia-renewable-energy (accessed on 25 July 2021).
81. Kuchler, M.; Bridge, G. Down the black hole: Sustaining national socio-technical imaginaries of coal in Poland. *Energy Res. Soc. Sci.* **2018**, *41*, 136–147. [CrossRef]
82. Walczak, E.; Zaborowski, M. Energy Efficiency in Poland. 2017 Review. The Institute of Environmental Economics (IEE). Available online: https://depot.ceon.pl/handle/123456789/15592 (accessed on 24 May 2021).
83. Główny Urząd Statystyczny. Energy 2019. Available online: https://stat.gov.pl/obszary-tematyczne/srodowisko-energia/ (accessed on 24 May 2021).
84. The Potential for Investment in Energy Efficiency through Financial Instruments in the European Union. Poland In-Depth Analysis. Available online: https://www.fi-compass.eu/erdf/potential-investment-energy-efficiency-through-financial-instruments-european-union/poland (accessed on 6 August 2021).
85. Energy Policies of IEA Countries Poland 2016 Review. Available online: https://euagenda.eu/upload/publications/untitled-69050-ea.pdf (accessed on 24 May 2021).
86. IEA, Poland. Available online: https://www.iea.org/countries/poland/ (accessed on 24 May 2021).
87. Vehmas, J.; Kaivo-oja, J.; Luukkanen, J. Energy efficiency as a driver of total primary energy supply in the EU-28 countries—Incremental decomposition analysis. *Heliyon* **2018**, *4*, 1–23. [CrossRef] [PubMed]
88. Zeleňáková, M.; Fijko, R.; Diaconu, D.; Remeňáková, I. Environmental Impact of Small Hydro Power Plant—A Case Study. *Environments* **2018**, *5*, 12. [CrossRef]
89. Energy Consumption in the EU Increased by 1% in 2017. Available online: https://ec.europa.eu/eurostat/documents/2995521/9549144/8-07022019-AP-EN.pdf/4a5fe0b1-c20f-46f0-8184-e82b694ad492 (accessed on 24 May 2021).
90. EC Energy Topics. Available online: https://ec.europa.eu/energy/en/topics/energy-strategy-and-energy-union/governance-energy-union (accessed on 24 May 2021).
91. Enterdata. Hungary Energy Information. Available online: https://www.enerdata.net/estore/energy-market/hungary/ (accessed on 25 July 2021).
92. Odyssee-Mure. Hungary Profile. Available online: https://www.odyssee-mure.eu/publications/efficiency-trends-policies-profiles/hungary.html (accessed on 25 July 2021).
93. European Environment Agency Tracking Progress towards Europe's Climate and Energy Targets—Trends and Projections in Europe 2018. 2018. Available online: https://www.eea.europa.eu/publications/trends-and-projections-in-europe-2018-climate-and-energy (accessed on 24 May 2021).
94. European Union Scaling up Climate Action. Key Opportunities for Transitioning to a Zero Emissions Society. Full Report. 2018. Available online: https://climateactiontracker.org/documents/504/CAT_2018-12-06_ScalingUp_EU_ExecSum.pdf (accessed on 24 May 2021).
95. Energy Policies of IEA Countries—Slovak Republic 2018 Review. Available online: https://www.connaissancedesenergies.org/sites/default/files/pdf-actualites/energy_policies_of_iea_countries_slovak_republic_2018_review.pdf (accessed on 25 July 2021).
96. A Spotlight on Renewables in the Slovak Republic. Available online: https://www.iea-shc.org/Data/Sites/1/publications/2020-07-Country-Highlight--Slovak-Republic.pdf (accessed on 25 July 2021).
97. Auctions for the Support of Renewable Energy in Slovakia. Available online: http://aures2project.eu/wp-content/uploads/2020/09/AURES_II_case_study_planned_Slovakia.pdf (accessed on 25 July 2021).
98. Climate Action Tracker Partners Climate Action Tracker. 2015. Available online: https://climateactiontracker.org/ (accessed on 24 May 2021).

99. Klugman, C. The EU, a World Leader in Fighting Climate Change. European Parliamentary Research Service. 2018. PE 621.818. Available online: https://www.europarl.europa.eu/thinktank/en/document.html?reference=EPRS_BRI(2018)621818 (accessed on 24 May 2021).
100. ArcelorMittal Wyłączy Wielki Piec w Krakowskiej Hucie. Trzy Scenariusze dla Pracowników. (ArcelorMittal Will Shut down the Blast Furnace at the Kraków Steelworks). Available online: https://poland.shafaqna.com/PL/AL/1052383 (accessed on 24 May 2021).

Article

Current (2020) and Long-Term (2035 and 2050) Sustainable Potentials of Wood Fuel in Switzerland

Matthias Erni *, Vanessa Burg, Leo Bont, Oliver Thees, Marco Ferretti, Golo Stadelmann and Janine Schweier

Federal Institute for Forest, Snow and Landscape Research (WSL), Zürcherstrasse 111, CH-8903 Birmensdorf, Switzerland; vanessa.burg@wsl.ch (V.B.); leo.bont@wsl.ch (L.B.); oliver.thees@wsl.ch (O.T.); marco.ferretti@wsl.ch (M.F.); golo.stadelmann@wsl.ch (G.S.); janine.schweier@wsl.ch (J.S.)

* Correspondence: matthias.erni@wsl.ch; Tel.: +41-44-739-21-24

Received: 28 September 2020; Accepted: 12 November 2020; Published: 23 November 2020

Abstract: Wood fuel has become central in environmental policy and decision-making processes in cross-sectoral areas. Proper consideration of different types of woody biomass is fundamental in forming energy transition and decarbonization strategies. We quantified the development of theoretical (TPs) and sustainable (SPs) potentials of wood fuel from forests, trees outside forests, wood residues and waste wood in Switzerland for 2020, 2035 and 2050. Ecological and economic restrictions, timber market situations and drivers of future developments (area size, tree growth, wood characteristics, population growth, exporting/importing (waste wood)) were considered. We estimated a SP of wood fuel between 26.5 and 77.8 PJ/a during the three time points. Results demonstrate that the SP of wood fuel could be significantly increased already in the short term. This, as a moderate stock reduction (MSR) strategy in forests, can lead to large surpluses in SPs compared to the wood fuel already used today (~36 PJ/a), with values higher by 51% (+18.2 PJ) in 2020 and by 59% (+21.3 PJ) in 2035. To implement these surpluses (e.g., with a cascade approach), a more circular economy with sufficient processing capacities of the subsequent timber industries and the energy plants to convert the resources is required.

Keywords: bioenergy; energy transition; forest management strategy; potential analysis; woody biomass; wood fuel

---

1. Introduction

In several European countries, an interest in using natural renewable resources to support the transition to clean energy has risen [1] (e.g., [2–4]). Biomass, in particular, is seen as a promising source for renewable energy and its use could cover a significant share of the primary energy consumption by 2050 [5,6], thus contributing to the achievement of European decarbonization goals. Consequently, the use of woody biomass has received attention but also raised several concerns [7]. While some of these concerns are related to the low energy density and conversion efficiency of wood fuel compared to fossil energies, as well as externalities such as the impact on air quality (with effects on human health) [7,8], the issue of possible conflicts among the multiple goods and services provided by forests that an augmented demand of wood for bioenergy may cause is central.

Forests provide a wide range of goods and services and are relevant for policy and decision-making processes in cross-sectoral areas [9], and are under pressure due to biotic and abiotic agents, whose action is exacerbated by events related to climate change, such as the 2018 megadrought (e.g., [10]). Therefore, some tension exists between "demands for more intensive management of biomass from forests and the contributions made by the same biomass in situ to soil fertility, biodiversity and protective functions" [7]. Sustainable and climate-smart forest management (e.g., [11,12]), and the cascade use of wood in long-lived products combined with end utilization for energy production [13–15], represent two necessary prerequisites for any sustainable and climate-friendly use of woody biomass as bioenergy (hereafter referred to as wood fuel). Thus, wood may only be used for energy production at the end of its life cycle or when it is not suitable for further material processing.

Switzerland has a long tradition of sustainable forest management [16], with forest areas and growing stocks continuously increasing over the past 40 years [17]. In this context, the sustainable use of wood fuel from forests can further help in reducing carbon dioxide (CO_2) emissions. It offers good opportunities to substitute energy and emission-intensive energy carriers and materials [18]. Moreover, it may support energy security and decarbonization strategies, while at the same time addressing multiple societal issues, e.g., solutions for the transportation sector and independence from imports [19,20]. Substituted products could be used for important high-end applications, such as the clean energy transition [21,22], urgent medicine, and the cosmetics, textile, and hardware industries.

Since wood fuel is intended to favor decarbonization processes, it is important that the interaction of forests with climate change is considered properly. Although younger, faster-growing forests usually have a higher rate of carbon uptake, it is the older forests with long-rotation periods and protected old-growth forests that have the highest carbon stocks. A key point for the sustainable use of wood fuel from forests is therefore the "payback time", i.e., the number of years it takes to offset the carbon emissions generated by harvesting the forest and the emissions caused if used for bioenergy. From this perspective, by regulating stand density and the age of forests, forest management has a substantial role in ensuring sustainable and climate-smart use of forest wood resources. In order to balance out society's manifold demands on forestry [23,24], any estimates of biomass availability should consider not only the theoretical potential, i.e., the upper limit of a given type of woody biomass available at a certain point in time, but also the sustainable potential, i.e., the share of the theoretical potential that can actually be used when other constraints (e.g. ecological) on management approaches are accounted for [20].

Many Central European countries refrain from full-tree use and have restricted the removal of small cutting volumes (non-commercial wood, e.g., slash, leaves and needles) in order to maintain nutrient cycles, soil quality as well as biodiversity. Forming technical projections and aspirational goals for future bioenergy can be difficult [23] but they are necessary, as stakeholders need reliable data on the development of woody biomass in order to plan infrastructures required for implementing the use of woody and non-woody biomass for energy [25]. In multiple studies, regional [26], national [20,27–30] and international [5,6,31,32] potentials of woody and non-woody biomass for energy use have been estimated. Results show that its use could be significantly increased in many countries already in the short term [20,33] and will most likely result in multiple advantages (e.g., [15,34–37]). However, as the availability of woody resources is limited [1,6], especially for energy purposes [20], knowledge is urgently needed on their (current and projected) availability in quantitative terms as well as the environmental and economic consequences of their use for bioenergy. While future estimates for non-woody biomass types exist [38,39], available estimates of wood fuel biomass potentials refer to specific types of woody biomass only, e.g., the biomass potential from short rotation coppices [40–42], from forests and other landscapes in urbanized regions [42], as well as biogenic waste and residues [43]. Estimates of the future availability of all woody biomass types at the country level is still missing.

With this study, we aim to investigate the availability of woody biomass feedstock resources originating from forests, trees from outside forests, residues and waste wood that can be used for wood fuel today (2020) and in the long term (2035; 2050). For all woody biomass types, we describe the time horizons and methods (frame), the connections between different wood types, their use, the uncertainty and its effects on the potentials, and the influence of different key drivers (i.e., factors that may drive the future availability of a given biomass type) on future estimates. The following specific research questions are addressed:

(i)   What sustainable potential of woody biomass feedstocks can be used for wood fuel today (2020) in the long term (2035; 2050), and what is the impact on the key drivers to predict future amount of wood fuel?

(ii)   What are realistic ranges of wood fuel predictions under different (a) forest management strategies, (b) ecological restrictions (e.g., protected areas, harvest losses), (c) wood markets (material use, recycling/upcycling) and (d) costs?

(iii)   What are the consequences of additional wood fuel use (e.g., effects of different costs)?

We focus on Switzerland as a case study, using data from the Swiss National Forest Inventory (NFI) [17] and models for future harvesting, forest management and wood fluxes. Additionally, the established framework of the Swiss Competence Center for Bioenergy Research (SCCER BIOSWEET) provided valuable results on the use of biomass for energy. Bridging and extending these activities was a chance to assess the development of all woody biomass types for energy use with a nationwide methodologically identical approach.

2. Materials and Methods

We applied a conceptual framework that explicitly incorporates the upper limit of annual wood fuel use, resulting in the theoretical (TP) and the sustainable potential (SP). Both are determined by ecological and socio-economic restrictions and assessed for four woody biomass types (wood fuel from forests, trees from outside forests, wood residues, waste wood; cf. Section 2.2).

2.1. Derivation of Wood Fuel Potentials

To estimate the wood fuel potentials (i), we considered (ii) three forest management strategies and (iii) three scenarios with different intensities of wood fuel use. Due to the influence of wood markets on the share of harvested wood allocated to material use, the classification into assortments was determined using (iv) two wood market situations. Finally, the consideration of (v) biomass-specific key drivers enabled the estimation of wood fuel potentials not only for today (2020) but also for future points in time (2035 and 2050) (Figure 1). In detail:

(i)   Theoretical and sustainable wood fuel potentials. Biophysical constraints and forest management strategies largely regulate the increment and possible stock reductions in forests and other landscapes, resulting in the theoretical potentials (TPs), and thereby determine the upper limit of annual wood fuel use. By subtracting relevant ecological and socio-economic restrictions, the amounts of biomass available for wood fuel are reduced, resulting in the sustainable potentials (SPs). SP provides a more realistic estimation of the amount of wood fuel produced annually and is the consequence of balancing societies' multifunctional forestry mandates [20,23,44]. However, some restrictions have only indirect effects on wood residues and waste wood as they occur after timber harvesting (dashed line in Figure 1). For both TPs and SPs, the specific assumptions made for each woody biomass type are described in the respective sections. In the case of wood fuel derived from forests, we additionally determined the ecologically sustainable potential (ESP), which provides the upper limit if significantly lower costs, higher prices and consequently larger revenues are expected and mainly ecological constraints dominate (cf. Figure 2).

(ii)   *Forest management strategies.* Three forest management strategies were simulated by using MASSIMO [45], a management scenario simulation model to project growth, harvests and carbon dynamics of Swiss forests. It is based on empirical models that have been fitted with data from the Swiss NFI [46], and it includes HeProMo [47,48], a model that can estimate potential timber harvesting productivities and costs. Thus, it was possible to estimate the harvesting cost for each NFI plot based on the amount of harvested timber, harvesting methods and transport distance to the nearest forest road. It was initialized on 5086 NFI sample plots with productive forests, based on NFI2 (1983–1985) and NFI3 (2004–2006) surveys. As MASSIMO includes random components, we replicated the simulations 20 times. The version applied (NFI2/NFI3) does not take into account climate change but includes random components for storm damage, harvesting and single tree mortality [45], thereby taking into account climate-induced extreme events that can cause major damage and further impact the wood-processing industries and trades [49]. The following forest management strategies were calculated:

a. Business as usual, representing a *continuous growing stock increase* (CSI). It reflects the current harvesting and management practices in Switzerland and therefore can be seen as a reference scenario. It leads to increasing growing stocks in all Swiss regions except for the Swiss Plateau.

b. Medium-intensity management, representing a *moderate stock reduction* (MSR). It targets growing stocks of 300–310 m^3/ha until 2046 and is intended to accelerate growth by reducing the stock of the forests.

c. High-intensity management, representing a *large stock reduction* (LSR). It accounts for a high demand for wood fuel from forests until 2046, leading to more frequent thinning and 40% shorter rotations, with target growing stocks of 250 m^3/ha and a range of 200–300 m^3/ha.

(iii) *Level of wood fuel consumption.* Three scenarios of wood fuel use were applied to estimate realistic ranges of TPs and SPs. They were calculated for all analyzed types of biomass and covered the minimum, maximum and expected levels of wood fuel consumption. The largest deductions due to the restrictions (i) result in the minimum, the smallest in the maximum and the deductions of average restrictions in the expected potentials. Thereby, the approach accounts for uncertainties and diverse viewpoints regarding the potentials use (cf. Figure 2).

(iv) *Timber market situations.* Two timber market situations were considered, wood energy oriented (EO+) and less wood energy oriented (EO−). They differ in the sorting of the assortments and include the differentiation between wood for energy and for material uses (Supplementary Information S1, Table S5). The market situation EO+ represents a more energy-intensive use directly after harvesting. In contrast, EO− represents a more cascade-oriented market, where a greater share of wood is directed towards a preliminary material use and is available for energy as wood residues or waste wood at the end of the life cycle. Both EO− and EO+ were defined in coordination with the Swiss forestry and timber industry associations [50].

(v) *Key drivers.* The consideration of biomass-specific key drivers serves to project the wood fuel potentials of all analyzed woody biomass types by using them as development indicators of TP. We subtracted the relevant biomass-specific restrictions from future TPs to determine SPs. Detailed information on the biomass-specific drivers is given in the corresponding Sections 2.2.1–2.2.4. In some cases, information on the possible development of the present SPs was also available and was applied directly.

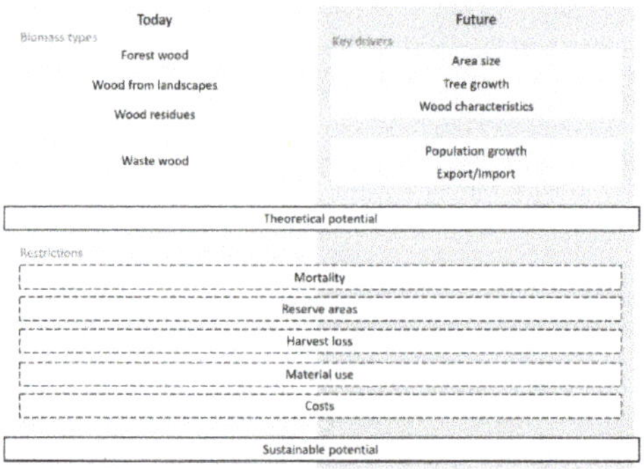

Figure 1. Procedure for the estimation of theoretical and sustainable wood fuel potentials (TPs and SPs) today (2020) and in the future (2035 and 2050).

Theoretical Potential (TP)
Determines the upper limit of annual wood fuel use of harvestable timber quantities, the accumulation of wood residues after processing, and the amounts of waste wood after a products' life cycle is complete.

Sustainable Potential (SP)
Ecological and socio-economic restrictions determine the differences between the TPs and SPs.

Ecological Sustainable Potential (ESP)
Determines the ecologically sustainable potential (ESP), which provides the upper limit if compared to the SP significantly lower costs, higher prices, and consequently larger revenues are expected and mainly ecological constraints dominate. This potential is calculated for wood fuel derived from forests only.

Scenario maximum, expected, minimum
For each of the potentials, three scenarios were calculated. These are the scenarios maximum, expected and minimum. The largest deductions due to the restrictions result in minimum, the smallest in maximum and the deductions of average restrictions in the expected potentials.

Figure 2. Definition of the different potentials and scenarios.

2.2. Assessment of Woody Biomass Types

Considering multiple utilization paths, the following four woody biomass types were considered relevant for wood fuel in Switzerland [30,51]:

(i) *Wood fuel from forests*, which is extracted from forests and used directly for energy purposes.

(ii) *Wood fuel from trees outside forests*, which is from trees and shrubs cut and pruned on landscapes or settlement areas outside forests.

(iii) *Wood fuel from wood residues*, which includes slabs, cuttings, sawdust and wood shavings that arise in the wood product manufacturing and processing industries [52]. These residues are mainly a by-product of saw-, planing- and gluing mills, carpentries and joineries within the woodworking industry. The potentials depend on (i) the amount of wood harvested in forests, and (ii) its assortments, which has the advantage of framing the potentials in relation to their original source.

(iv) *Wood fuel from waste wood*, which is defined as waste (after the material use of a wooden product) [53] and includes packaging made of wood, waste wood from building sites, demolitions, renovations and conversions, and shredded wood, including sieve overflow.

2.2.1. Wood Fuel from Forests

Applied models. By using MASSIMO [45] and the integrated HeProMo model [47,48], forest development, future growth, costs and harvesting potentials were simulated with time steps of 10 years, taking random storm damage, harvesting and single tree mortality into account (e.g., [45,54,55]).

Considered forest management strategies. All forest management strategies (CSI, MSR, LSR) and cost categories were described and applied according to Section 2.1.

Calculation of theoretical potentials. The TPs (TP_{2020}, TP_{2035}, TP_{2050}) correspond to the annual increments including the amounts of stock reductions. To consider the range of the potentials for each management strategy, the standard errors of the MASSIMO outputs for increment and the eventual stock reduction were applied (Table 1). The future TPs were based on the MASSIMO results for the periods 2027–2036 (TP_{2035}) and 2047–2056 (TP_{2050}). As MASSIMO assumes mortality rates of 15% for CSI and MSR and 10% for LSR, mortality rates were re-added for the TP because trees that died could have been used for energy before decomposition.

Table 1. Standard error of the increment derived from MASSIMO for the different time perspectives for Switzerland.

Time/Management Strategy	CSI	MSR	LSR
2020	2.6%	3.0%	3.4%
2035	9.2%	3.4%	3.4%
2050	11.7%	5.3%	2.9%

CSI = continuous growing stock increase, MSR = moderate stock reduction, and LSR = large stock reduction.

Key drivers considered in the future TPs. We used MASSIMO's standard implementations with a storm periodicity of 15 years, as well as the key drivers forest area size, forest growth and wood characteristics, all identified by literature review. Forest area size: Based on [56,57], we assumed a forest area increase of 0.19% per year, which resulted in a summed up forest area increase of almost 3% for TP_{2035} and 6% for TP_{2050} compared to TP_{2020}. Forest growth: While factors such as climate change and N deposition may negatively impact tree growth in the lowlands [58,59], gross growth may increase at higher altitude [60,61]. Based on [49,62] and the higher forest cover in the Alps than in the lowlands, we assumed that both effects result in an increasing TP of 11% for TP_{2035} and 10% for TP_{2050} compared to TP_{2020} (7.9 m^3/ha [63]). Wood characteristics: MASSIMO outputs result in different proportions of broadleaves and conifers over time. This shift in species influences the calorific value and thus the energy content of the wood. In MASSIMO, the growth of small trees (below the calliper threshold) is assumed to follow the same tree species-specific growth models as trees beyond the calliper threshold of 12 cm diameter at breast height. This assumption facilitates the simulation of fast-growing and shade-tolerant species in the regeneration, slightly tending to an overestimation of conifers [55] (for more details on regeneration in MASSSIMO, cf. [45]).

Calculation of sustainable potentials. To determine the SPs of wood fuel from forests, we considered restrictions with respect to mortality, reserve areas, harvest losses, material use and costs in the named order [20]. Mortality: The share of mortality was subtracted according to the assumptions made in the respective management strategies (CSI, MSR: 15% of increment; LSR: 10%). Reserve areas: Based on [64], we considered todays forest reserve areas (4.8%) when estimating SP_{2020}. To reflect the minimum that policy aims for [65], we doubled the extend of reserve areas (~10%) for both SP_{2035} and SP_{2050}. Harvest losses: Based on data from [26], we calculated standard errors of harvest losses and estimated the minimum SPs by subtracting the largest possible harvest losses and the maximum potential, respectively (Table 2). Material uses: Based on [20], we applied the two wood market situations "less wood energy oriented" (EO−) and "wood energy oriented" (EO+) for the sorting of assortments into energy and material wood use (Supplementary Information S1, Table S5). Costs: In addition to the harvesting costs divided into 25 CHF cost classes, costs for chipping and transportation were considered (40 CHF/m^3). For SP_{2020}, the threshold for economically acceptable harvesting, chipping and transportation costs was set, based on [20], at today's market price of 120 CHF/m^3 for wood production areas and 190 CHF/m^3 for protection forest areas, accounting for subsidies given for the management of protection forests. To account for the minimum and maximum potentials we assumed ±10 CHF/m^3 of the total costs.

Table 2. Range of harvest losses (minimum and maximum).

Harvest Losses	Conifers		Broadleaves	
	Merchantable	Brushwood	Merchantable	Brushwood
Minimum wood fuel SPs	9%	61%	16%	55%
Maximum wood fuel SPs	7%	55%	10%	45%

Calculation of ecologically sustainable potentials (ESPs). In addition to TP and SP, we determined the situation of the sustainable potential for wood fuel from forests when costs are neglected. Therefore, cost restrictions were not subtracted from TPs. This results in the ESP, which considers ecological restrictions only and is in line with [20].

2.2.2. Wood Fuel from Trees Outside Forests

Applied models. A model was established based on geographic information systems (GIS), calculating wood growth per surface using a 100 m × 100 m grid of Switzerland [51].

Considered forest management strategies. None, as the procedure estimates wood from trees outside forests.

Calculation of theoretical potentials. To estimate TPs, an increment was assigned to the 28 identified wood-stocked categories outside forests (Supplementary Information S2, Table S6) shown in the areal statistics [66]. Increment accounts for site characteristics, such as precipitation, soil conditions and exposure of the trees [67]. Specifically, we assumed optimal growth of each stocked category based on literature review and expert surveys (Supplementary Information S2, Table S7). In a next step, we corrected optimal growth values with realistic coverage levels (Supplementary Information S2, Table S7, column 6) and applied the dependence of increment on altitude [63] using net growth factors (Supplementary Information S2, Table S8). Therefore, we normalized the net growth. The altitude class with the highest growth value of each region was assigned "1", while net growth factors were assigned proportionally smaller values for the others according to the growth measurements of the NFI [63]. Afterwards, net growth factors (Supplementary Information S2, Table S8) were assigned to all of the 4.1 million georeferenced sample points of the areal statistics [66]. The results were then joined with the digital elevation model of Switzerland [68], thereby assigning an altitude to each of the areal statistics sample points. Finally, the combination of the stocked areas outside forests with altitude and the corresponding net growth factors (Supplementary Information S2, Table S8, lower part) made it possible to estimate net growth in $[m^3/ha*a]$ for each of the georeferenced sample points of the areal statistics. Thus, we multiplied the growth of realistic coverage levels of the 28 wood-stocked categories (Supplementary Information S2, Table S7) with their net growth factors while also accounting for their location and altitude (Supplementary Information S2, Table S8). When summed over all of Switzerland this equals TP.

Key drivers considered in the future TPs. Increment: For wood from trees outside forests, we made the same assumptions as for wood fuel from forests, i.e., an additional 11% woody biomass available for the TP_{2035} and 10% for the TP_{2050} compared to the TP_{2020} (Section 2.2.1). *Wood characteristics:* The wood from landscape maintenance was assumed to be 90% broadleaves for the TP_{2020} [51]. As this percentage is quite high, it was assumed to stay at 90% for TP_{2035} and TP_{2050}. *Areal changes:* To quantify the wood-stocked areas outside forests for TP_{2035} and the TP_{2050}, we linearly extrapolated the past development based on the areal statistics from 1979/1985 and 2004/2009 [66]. The areas increased by approximately 12% between these two periods, which corresponds to an annual average increase of 0.46%. This total increase was distributed equally across all types of wood from trees outside forests until 2035. However, stagnation in the year 2035 was assumed.

Calculation of sustainable potentials. Based on the results from the TPs, restrictions were considered that apply when technical-economic and socio-political requirements were included. In particular, harvesting, chipping, and transportation costs were important, and all of which largely depend on the accessibility of the area. *Technical-economic accessibility:* The distance of the landscape areas to the road network [69] is important when considering forwarding distances. Therefore, we considered the distances between the wood-stocked areas on landscapes and the nearest road (>2 m width). In cases where the distance was <50 m, 100% of the wood was considered for the TP, whereas 60% of the TP was considered where the distance was 50–150 m and no wood was considered where the distance was >150 m. This is in line with [51,67]. *Ecological socio-political accessibility:* As harvesting in steep terrain is avoided whenever possible, for worker safety reasons and prevention of erosion, the slope was calculated with the digital elevation model [68]. In general, we assumed that sites with a slope gradient of more than 70% were harvested only if necessary for safety reasons (e.g., better view along roads or in urban areas, Table 3).

Key drivers considered in the future SPs. Areal changes: The annual increase of 0.46% from former areal statistics was applied (cf. key drivers in the future TPs). As for the TPs, we assumed this increase would persist up to the year 2035 (SP_{2035}), resulting in an additional 12% that was distributed equally across on all stocked categories of trees outside forests (Supplementary Information S2, Table S7). *Availability:* We assumed similar road networks and forwarding distances as for SP_{2020} for future points in time (SP_{2035} and SP_{2050}). We also assumed that there were no changes in topography or slopes within these timeframes. *Uncertainty:* We chose high uncertainty values of 20% for today (2020), 40% for 2035 and 60% for 2050 (for more details on the technical assessment, cf. [51]).

Table 3. Technical limitations due to slopes.

Categories of Wood from Landscapes	Modelling	Notes
Built-up areas	All the areas are considered for the calculation of SPs	Maintenance and management due to safety aspects
Transportation areas		
Agricultural areas, tree and brush vegetation	Only areas with a slope <70% were considered	Safety aspects not relevant
Groves, hedges Embankments (of rivers and lakes)		

2.2.3. Wood Fuel from Wood Residues

Applied models. Wood harvesting potential. (i) We applied the models *MASSIMO* and *HeProMo* (cf. Section 2.1) and used the same approach as for "wood fuel from forests" (Section 2.2.1), neglecting the assortment (market situations; EO−, EO+) because material, wood-processing and energy uses are part of the model *"wood fluxes Switzerland"*. (ii) This model is based on [70] and includes the quantification of wood residues available for fuel as all wood residues occurring along the value chain of the wood-processing and woodworking industries (Figure 3). The model is divided into three main pillars (wood energy, stem wood and industrial timber). For the determination of wood residues for wood fuel, we summed up wood residues intended for energy use (Figure 3, framed in red). *Linkage of the two models.* (i), (ii) Wood harvesting SPs and TPs from (i) were used as possible wood harvest (cf. wood harvesting potential, Figure 3) input in model (ii) for each point in time (2020, 2035, and 2050). This combination allowed the distribution of wood harvesting TPs and SPs (i) based on five-year average values of the wood fluxes in Switzerland (ii) (Figure 3) [70].

Considered forest management strategies. The moderate stock reduction (MSR) strategy was used, as it is expected to be favored by the forestry and timber industry [20]. Additionally, it allows temporary surpluses for the SP_{2035} and SP_{2050} coming from increased harvests due to stock reductions until 2046, while not undercutting the SP_{2050} coming from the CSI (business as usual) strategy and still providing compartments for wood processing.

Calculation of theoretical potentials. TPs consider all wood residues to be used directly for energy. This means that parts used for further material production and unidentifiable quantities in terms of energy versus material use (Figure 3, dashed in red) are entirely attributed to the TPs (cf. applied models).

Key drivers considered in the TPs. As TPs of wood fuel from forests (Section 2.2.1) are similar to TPs of wood harvesting potentials (i) (cf. applied models), we considered no further drivers.

Calculation of sustainable potentials. SPs consider all fluxes of wood residues that are no longer used for reprocessing (Figure 3, framed red). Concerning wood harvesting SPs, the restrictions of mortality, reserve areas, harvest losses and costs were considered in line with the procedure used for wood fuel from forests (Table 4, respectively Section 2.2.1).

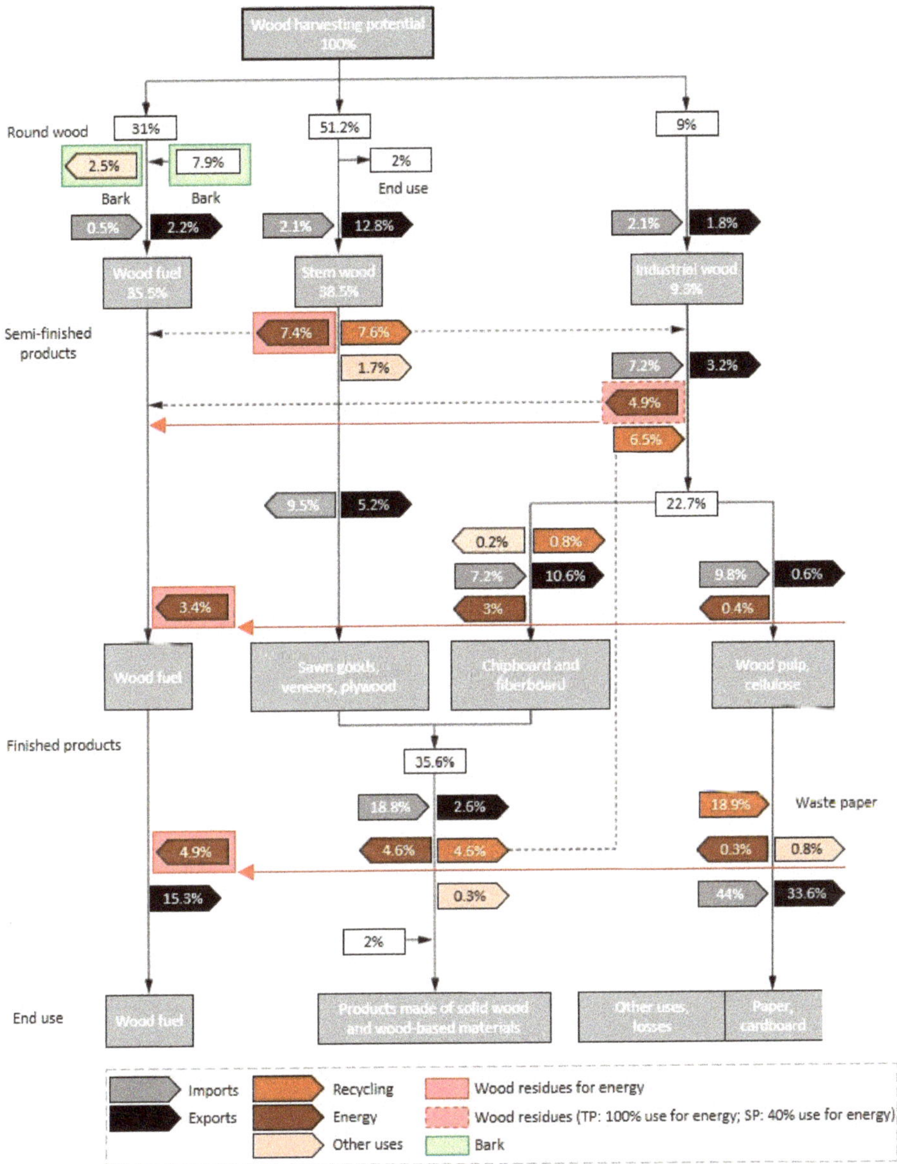

Figure 3. Wood fluxes in Switzerland ([70] modified, 2010-2014, five year average). Shown are the shares of the wood residues for recycling and energy in relation to the forest wood harvesting potential. The quantities of wood residues used to generate energy are framed in red. Bark (framed in green) was not part of the wood residues estimation, as its original source is in the forests (Section 2.2.1). It corresponds to approximately 8% of the forest wood utilization potential.

Table 4. Minimum, expected and maximum restrictions for the calculation of sustainable forest wood potentials.

Restriction	Specification	Minimum	Expected	Maximum
Mortality [%]			-/15/-	
Reserve areas [%]	SP_{2020}	~10.0	4.8	4.8
	2035 and 2050	~10.0	~10.0	~10.0
	$SP_{2035/2050}$	~10.0	~10.0	~10.0
Harvest losses [%]	Conifers brushwood (incl. bark)	61	58	55
	Conifers merchantable	9	8	7
	Broadleaves brushwood (incl. bark)	55	50	45
	Broadleaves merchantable	16	13	10
Costs [CHF/m^3]	Wood production forests	110.0	120.0	130.0
	Protection forests	180.0	190.0	200.0

2.2.4. Wood Fuel from Waste Wood

Applied models. A survey of all waste wood disposal companies and waste wood transporters within Switzerland was conducted [71] to quantify the amount of waste wood and its potentials for energy use. The survey was based on [72] and was adapted and implemented for Switzerland. In total, 295 companies reported about their activities in the transport and disposal of waste wood in 2014. Based on this survey, data for 154 transportation and disposal companies that did not respond were projected. This provided the basis for the estimation of the waste wood energy SPs and TPs (for details, cf. [71]).

Considered forest management strategies. As the estimation of waste wood potentials is based on an actual survey, the situation is comparable to the strategy CSI with a considerable time lag because waste wood occurs at the end of the life cycle. However, as many semi-finished and finished products are imported [73], the domestic production has minor importance.

Calculation of theoretical potentials. The TPs correspond to the total market volume. The market volume reflects the maximum quantity of waste wood available on the market, including both domestic use and export. It contains the quantities used within the country and the quantities sold abroad to end users. Today, 0.3 M t (0.56 M m^3) of waste wood, or 31% of the market volume, is exported [71]. This is important because exported waste wood could be used domestically.

Calculation of sustainable potentials. The SPs reflect the cascading (material use) as well as the appropriate disposal of contaminated waste wood inland and abroad. We determined the amount of waste wood already used for energy production in Switzerland and added the amount exported and used for energy production abroad (15% of the market volume or 0.15 M t (~0.28 M m^3); [71]).

Key drivers considered in the TPs and SPs. It was assumed that a growing population will increase the demand for wooden products (e.g., furniture) and, consequently, will produce more waste wood. Therefore, we accounted for population growth: based on [74], Switzerland's population (8.3 M permanent residents) is expected to increase by 19% by 2035 (approx. 1% annually) and by 23% by 2050 (approx. 0.76% annually), with deviations in these scenarios of 5%, and 10%, respectively. As other investigations of the waste wood already used for energy [75,76] show results with variations of 10% for the potential, we added an additional 10% uncertainty to our results.

2.2.5. Calculation of Energy Contents

General remarks. Volume- or weight-based data on wood harvests from forests and landscapes, wood residues, or wood wastes were converted into energy content (E) (Joules [J]). Depending on the biomass type, different values for the energy content per volume or weight were applied. If not indicated otherwise in the respective sections, the following values were used. The energy content of fresh forest wood is 2260 kWh/t for conifers and 2160 kWh/t for broadleaves [77]. For the conversion of kWh/t into J, we used the factor 3.6×10^6 [78]. For fresh wood from forests, we applied volumes of 0.758 t/m^3 for conifers and 1.116 t/m^3 for broadleaves. For air-dry wood, we applied volumes of 0.52 t/m^3 for conifers and 0.74 t/m^3 for broadleaves [51].

Wood fuel from forests. The energy content is given for fresh wood of conifers and broadleaves:

$$E_{conifers}[J] = volume_{conifers}[m^3] * 0.758\left[\frac{t}{m^3}\right] * 2260\left[\frac{kWh}{t}\right] * 3.6 * 10^6\left[\frac{J}{kWh}\right]$$

$$E_{broadleaves}[J] = volume_{broadleaves}[m^3] * 1.116\left[\frac{t}{m^3}\right] * 2160\left[\frac{kWh}{t}\right] * 3.6 * 10^6\left[\frac{J}{kWh}\right]$$

Wood fuel from trees outside forests was estimated based on weight. The energy content was given for fresh wood of conifers and broadleaves. The ratio of conifers to broadleaves was assumed to be 1/10 (cf. Section 2.2.2).

$$E_{conifers}[J] = weight_{conifers}[t] * 2260\left[\frac{kWh}{t}\right] * 3.6 * 10^6\left[\frac{J}{kWh}\right]$$

$$E_{broadleaves}[J] = weight_{broadleaves}[t] * 2160\left[\frac{kWh}{t}\right] * 3.6 * 10^6\left[\frac{J}{kWh}\right]$$

Wood fuel from wood residues. The wood used for energy directly after production processes in sawmills or energy plants was assumed to have a high moisture content (50%), whereas residues from further processing were assumed to be air dry. For TPs, 66% of the wood residues are freshly used and 34% are used later on and therefore considered air dry. For SPs, 60% are freshly used, while 40% come from further processing (air dry) (based on Section 2.2.3, model "wood fluxes Switzerland"). The cutting of conifers and broadleaves occurs in a ratio of 20/1 (roundwood cutting ratio, sawmill residues) [79]. This ratio was also used for wood residues from woodworking companies.

For TPs:

$$E_{conifers}[J] = 0.66 * volume_{conifers}[m^3] * 0.758\left[\frac{t}{m^3}\right] * 2260\left[\frac{kWh}{t}\right] + 0.34$$
$$* volume_{conifers}[m^3] * 0.52\left[\frac{t}{m^3}\right] * 2260\left[\frac{kWh}{t}\right] * 3.6 * 10^6\left[\frac{J}{kWh}\right]$$

$$E_{broadleaves}[J] = 0.66 * volume_{broadleaves}[m^3] * 1.116\left[\frac{t}{m^3}\right] * 2160\left[\frac{kWh}{t}\right] + 0.34$$
$$* volume_{broadleaves}[m^3] * 0.74\left[\frac{t}{m^3}\right] * 2160\left[\frac{kWh}{t}\right] * 3.6 * 10^6\left[\frac{J}{kWh}\right]$$

For SPs:

$$E_{conifers}[J] = 0.60 * volume_{conifers}[m^3] * 0.758\left[\frac{t}{m^3}\right] * 2260\left[\frac{kWh}{t}\right] + 0.40$$
$$* volume_{conifers}[m^3] * 0.52\left[\frac{t}{m^3}\right] * 2260\left[\frac{kWh}{t}\right] * 3.6 * 10^6\left[\frac{J}{kWh}\right]$$

$$E_{broadleaves}[J] = 0.60 * volume_{broadleaves}[m^3] * 1.116\left[\frac{t}{m^3}\right] * 2160\left[\frac{kWh}{t}\right] + 0.40$$
$$* volume_{broadleaves}[m^3] * 0.74\left[\frac{t}{m^3}\right] * 2160\left[\frac{kWh}{t}\right] * 3.6 * 10^6\left[\frac{J}{kWh}\right]$$

Wood fuel from waste wood was estimated based on weight. The primary energy content was calculated with the calorific value for conifers of 4.02 kWh/kg for conifers and 3.86 kWh/kg for broadleaves (moisture content 20%) [77]. Significantly more coniferous wood is used for processing. Therefore, we used a conifer to broadleaf ratio of 5/1 [73].

$$E_{conifers} = 0.8 \underset{of\ conifers}{_{for\ proportion}} * weight_{waste\ wood}[kg] * 4.02\left[\frac{kWh}{kg}\right] * 3.6 * 10^6\left[\frac{J}{kWh}\right]$$

$$E_{broadleaves} = 0.2 \underset{\substack{for\ proportion \\ of\ broadleaves}}{} * weight_{waste\ wood}[kg] * 3.86 \left[\frac{kWh}{kg}\right] * 3.6 * 10^6 \left[\frac{J}{kWh}\right]$$

3. Results

3.1. Wood Fuel from Forests

TPs and SPs show large differences in wood fuel from forests. Considering the different forest management strategies, the respective scenarios, and the reference year of the availability, from 10% up to a considerable 45% of SPs are available from TPs for direct energy use only.

3.1.1. Theoretical Potential

The TP of wood fuel from forests is between 75 and 132 PJ/a for the next 30 years, depending on the management strategy and reference year. Today and in 2035, the strategy LSR offers the greatest potential. Between 2035 and 2050, however, the volume is expected to decrease significantly and the potentials of the strategies CSI and MSR will exceed the potential of the strategy LSR in the long term (Figure 4; Supplementary Information S1, Table S1).

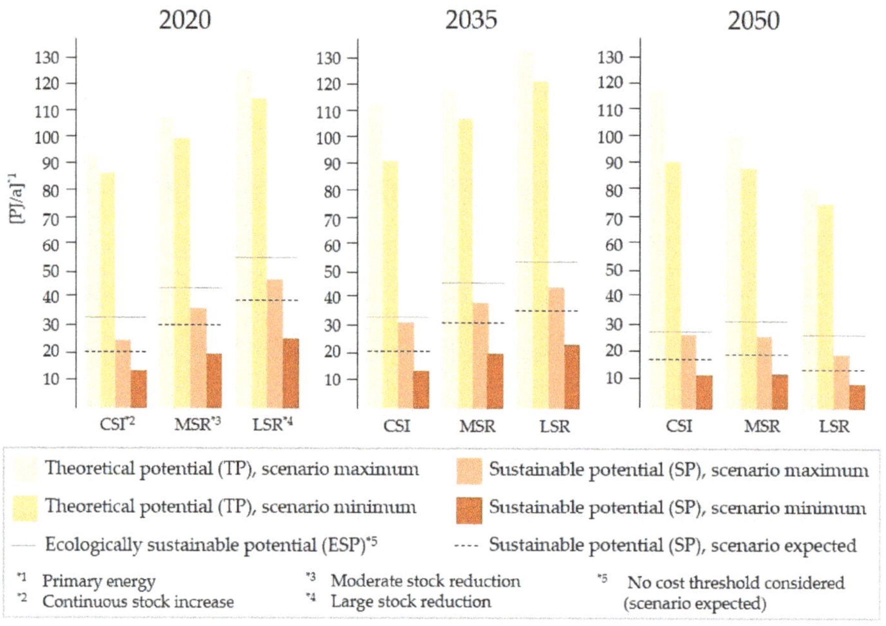

Figure 4. Development of the theoretical and sustainable potential (minimum and maximum) of wood fuel from forests for energy use in Switzerland from 2020 to 2050.

3.1.2. Sustainable Potential

The SP of wood fuel from forests is between 9.1 and almost 47.3 PJ/a for the next 30 years, depending on management strategy, timber market situation, the point in time examined and whether deductions are considered for mortality, reserve areas, harvest losses, material uses and wood harvesting, as well as chipping and transportation costs (Figure 4; Supplementary Information S1, Table S2). This represents a wide range of wood fuel that might be available from forests. The management strategy LSR leads to the largest SP in 2020 and 2035. It decreases, however, by 2035 and 2050 (Figure 4). Wood fuel potentials from the management strategies CSI and MSR will exceed those from LSR in the long term (Figure 4; Supplementary Information S1, Table S2; year 2050). The lower boundaries of the potentials are still within the range of those corresponding to LSR (Figure 4; CSI, MSR). This is mainly due to the uncertainty range given by the multiple replications of the scenario calculations (standard error), harvest losses, material uses of the wood (EO+ wood energy oriented vs. EO− less wood energy oriented) and costs or revenues of $\pm 10\,\mathrm{CHF/m^3}$ ($= 9.5\,€/\mathrm{m^3}$ or $10.3\,\mathrm{USD/m^3}$, actual exchange rate on 13.04.2020).

Neglecting costs results in the ecologically sustainable potential (ESP; cf. Section 2.2.1) leads to considerable surpluses (Figure 4 (grey lines) and Supplementary Information S1, Table S3). Comparing the ESPs of the different forest management strategies with the corresponding SPs (considering costs) of the CSI (business as usual) indicates surpluses between 64% and almost 200% for the minimum, 60% and 212% for the expected, and 32% and 148% for the maximum scenarios (Supplementary Information S1, Table S4).

3.2. Wood Fuel from Trees Outside Forests

Considering the respective scenarios and time points, TPs and SPs of wood fuel from trees outside forests can be expected to have only slightly higher availabilities in the future. Between 50% and 52% of SPs are available from TPs.

3.2.1. Theoretical Potential

The TP of wood fuel from trees outside forests is between 9.4 and 11.7 PJ/a, for the next 30 years until 2050 (Figure 5; Supplementary Information S2, Table S6). The wood fuel potential increases slightly until 2035 and then remains at the same level until 2050. This goes along with the higher growth increments of the individual trees expected due to climate change and the ingrowth of non-woody landscape areas.

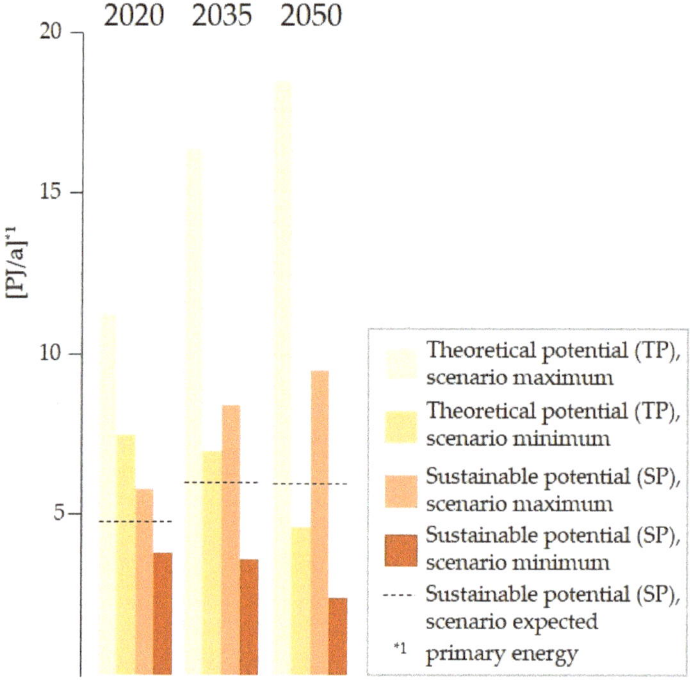

Figure 5. Development of the theoretical and sustainable potential of wood fuel from trees outside forests in Switzerland from 2020 to 2050.

3.2.2. Sustainable Potential

The SP of wood fuel from landscapes is between 4.8 and 6.0 PJ/a, for the next 30 years for the expected scenario (Figure 5; Supplementary Information S2, Table S6). The potential increases slightly until 2035 and then remains at the same level until 2050.

3.3. Wood Fuel from Wood Residues

TPs and SPs show very large differences in wood fuel from wood residues. Considering the respective scenarios and the time point of the availability, only 4–14% of SPs are available from TPs for energy use. Along the value chain of wood harvesting, SPs and TPs result in wood residues for energy of 37% (SPs) to 47% (TPs). At the semi-finished product level, wood residues for energy arise from processing industrial wood, mainly by the production of chipboard and fiberboard (SPs: 17%; TPs: 35%). The difference between TP and SP occurs because an optimistic cascade of use is applied to SPs, as recommended by [13–15]. Wood residues for fuel resulting from the production of solid wood and wood-based materials have a share of 21% (TPs) to 26% (SPs) and include sawn timber, veneer and plywood. Further, small quantities of wood residues occur as cellulose and wood pulp in the paper and cardboard industry.

3.3.1. Theoretical Potential

The TP of wood fuel from wood residues is between 53.9 and 68.8 PJ/a, for the next 30 years until 2050, depending on the standard error of the forest management strategy MSR, and the reference year (Figure 6; Supplementary Information S3, Table S9) and current wood market conditions (Section 2.2.3). The TP (Figure 6; Supplementary Information S3, Table S9) shows the upper limit of wood residues for energy use, if the total TP of forest wood is exploited for material use.

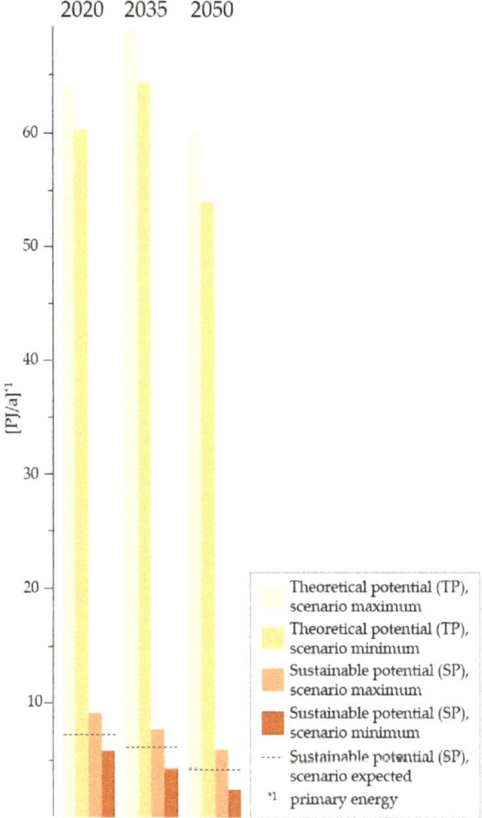

Figure 6. Development of the theoretical and sustainable potential of wood fuel from wood residues for energy use in Switzerland from 2020 to 2050.

3.3.2. Sustainable Potential

The SP of wood fuel from wood residues is between 2.4 and 9.1 PJ/a, for the next 30 years until 2050, depending on the forest harvests in reserve areas, costs and the reference year (Figure 6; Supplementary Information S3, Table S9). The SP represents a wide range of wood fuel from forests being available each year, also considering the wood market situation. The SPs of wood residues continuously decrease until 2050 to a minimum of 2.4 PJ/a and a maximum of 5.9 PJ/a. The forest wood harvesting potentials were used as input for the model (cf. "wood fluxes of Switzerland", Section 2.2.3) and are shown in Supplementary Information S3, Table S10.

3.4. Wood Fuel from Waste Wood

A large share of the TP of waste wood is sustainably available for wood fuel use. Considering the respective strategies and the time point of the availability, a considerable (82–84%) of SPs are available from TPs for energy use.

3.4.1. Theoretical Potential

The TP of wood fuel from waste wood ranges from 12.9 to 20.9 PJ/a, for the next 30 years (Supplementary Information S4, Table S11). At first, TP increases from 14.3 PJ/a (2020) to 16.5 PJ/a (2035) for the expected scenario. However, after 2035, it increases at a slower rate to 17.1 PJ/a until 2050.

Compared to today (TP$_{2020}$, expected scenario), in 2035 the minimum potential is higher by 3.5% and the maximum by 30%. In 2050, the minimum potential is 7% higher, while the maximum is 32% higher. Minimum quantities are 11% lower than the expected scenario and the maximum quantities are 23% higher in both 2035 and 2050 (Figure 7).

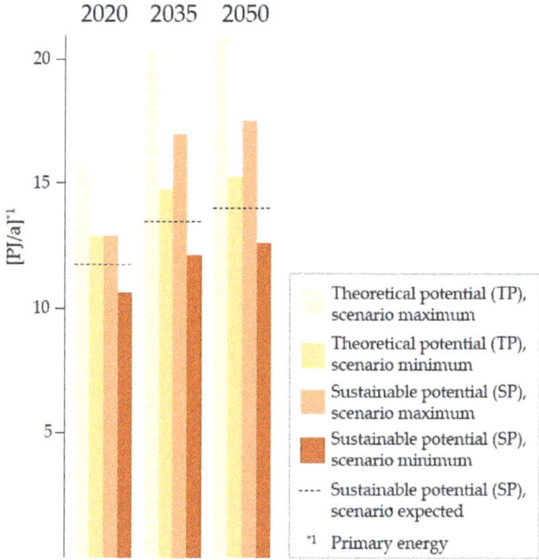

Figure 7. Development of the theoretical and sustainable potential of wood fuel from waste wood for energy use in Switzerland from 2020 to 2050.

3.4.2. Sustainable Potential

The SP of waste wood ranges from 10.6 to 17.5 PJ/a, for the next 30 years until 2050 (Figure 7 and Supplementary Information S4, Table S11). A slight increase in the potentials is expected. Compared to today (SP$_{2020}$), the minimum potential in 2035 is 3.5% higher, while the maximum potential is 19% higher. In 2050, the minimum potential is 19% higher and the maximal potential is 49% higher. Compared to the expected scenario, the minimum and maximum quantities are initially (in 2020) 10% lower and higher, respectively, the minimum potentials in 2035 and 2050 are 11% lower, and the maximum potentials are 26% higher in 2035 and 25% higher in 2050.

3.5. All Woody Biomass Types

The TP of wood fuel from all biomass sources is between 93.1 and 148.5 PJ/a, for the next 30 years until 2050, depending on the forest management strategy, forest policy implementation, exporting/importing and the reference year (Figure 8 and Supplementary Information S5, Table S12). The TP gives the upper limit of domestic wood for the energy use, not considering any material purposes. As wood residues and waste wood have their origins in forests and other landscapes, only these two primary sources need to be summed up to account for TP.

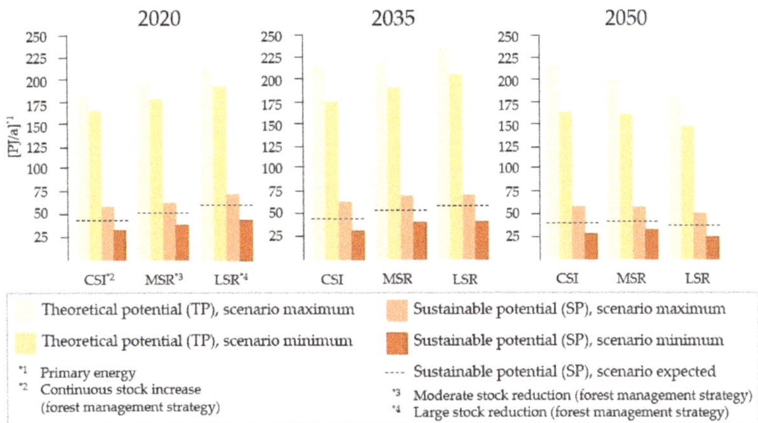

Figure 8. Development of the theoretical and sustainable potentials of all woody biomass types for wood fuel in Switzerland from 2020 to 2050. * Wood residues and waste wood have their origins in forest wood and wood from trees outside forests. Therefore, the TP of all domestic woody biomass consists only of the TPs of forest wood and wood from other landscapes. For detailed results, cf. Supplementary Information S5, Table S12.

The SP of wood fuel is between 26.5 and 77.8 PJ/a for the next 30 years until 2050, depending on the forest management strategy, development of costs, harvest losses, forest policy implementation (e.g., reserve areas, use of wood from landscapes), exporting/importing and the reference year (Figure 8, Supplementary Information S5, Table S13).

Compared to the SP_{2020} for the management strategy CSI, surpluses of 43% to 47% occur when the management strategy MSR is applied, while the implementation of LSR results in surpluses of 84–89% (Supplementary Information S1, Table S3) for 2020. For SP_{2035}, potentials 22% to 49% higher can be expected compared to continuation with the management strategy CSI. The large range can be explained mainly by the different sorting of wood compartments for material and energy use according to the EO− and EO+ market situations, the cost elasticity of 20 CHF/m^3, the different assumptions for the minimum and maximum harvest losses, and the larger deductions for reserve areas (cf. Section 2.2.1). For SP_{2050}, the potentials resulting from stock reduction strategies (MSR, LSR) decrease slightly compared to 2035. However, while potentials for MSR are within the same range as those for CSI, the potentials for LSR are lower, especially when considering an EO− market situation (Figure 8; Supplementary Information S5, Table S12).

Overall, the largest amount of wood fuel is provided from forests, but this share decreases continuously in the future. At the same time, the range of the amount of wood fuel from trees outside forests becomes broader. Potentials in the minimum scenario become lower and those in the maximum scenario become higher. Moreover, the share of wood residues decreases during the next 30 years, while the share of waste wood increases at least in the last time step (2035–2050) (Figure 9; Supplementary Information S5, Table S13).

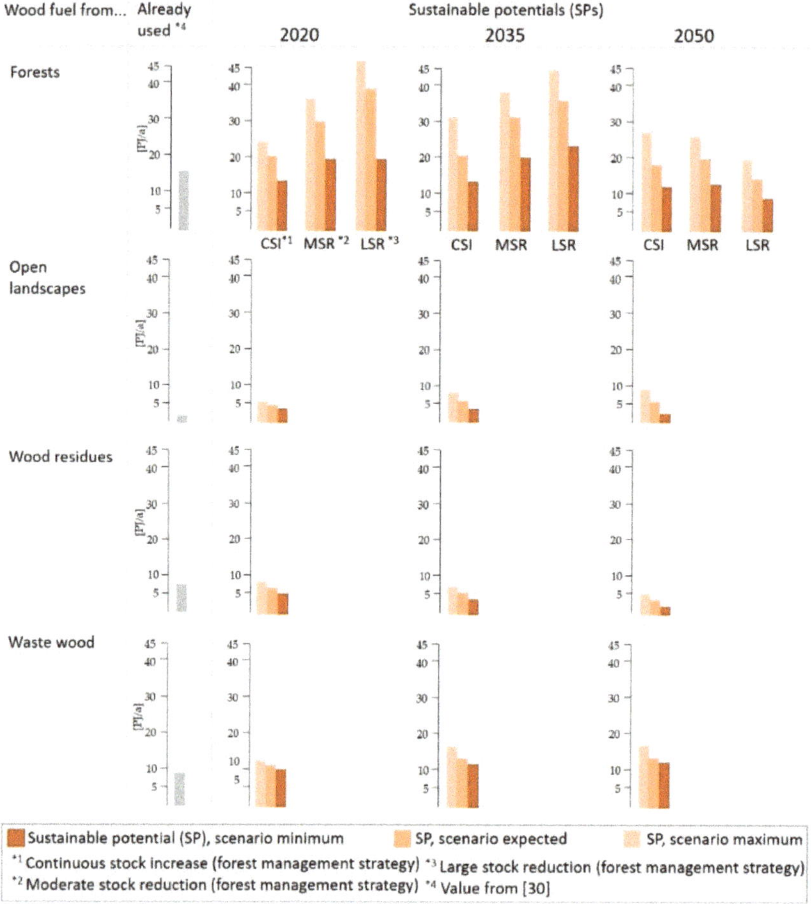

Figure 9. Development of the sustainable potentials (SPs) of wood fuel for all woody biomass types in Switzerland from 2020 to 2050. For detailed results, cf. Supplementary Information S5, Table S14.

4. Discussion

The current and long-term fuel potentials of woody biomass are compared with the annual gross energy consumption in Switzerland and with all biomass types (including non-woody) (Section 4.1). Since forest wood is the main source of all the woody biomass types, we analyze the potentials with regard to their original source but also to the subsequent timber industries (wood residues) and the final use of wooden products (waste wood) (Section 4.2). We differentiate between levels of wood fuel consumption by comparing the different assortments of wood compartments, which makes it possible to discuss the contribution to a more circular or cascade-oriented wood fuel market (Section 4.3). Finally, the key drivers are put into context of the wood fuel potentials and are discussed critically (Section 4.4).

4.1. Context of Wood Fuel Potentials

Today, the share of wood fuel is 3.8% of the total gross energy consumption in Switzerland (~1100 PJ) [80]. This corresponds closely to the SPs for the CSI (business as usual) strategy in 2020, according to the "expected" level of wood fuel consumption (CSI: 44.7 PJ, Figure 9, Supplementary

Information S5, Table S13). In the long-term perspective, the SP of CSI is rather stable (2035, 46.7 PJ; 2050, 42.6 PJ). Estimations for alternative, more intensive forest management strategies (MSR, LSR) show biomass surpluses in the short (2020) and mid-term (2035) compared with values for CSI. With these management strategies, the minimum SPs estimated for the wood consumption (SP_{2020} and SP_{2035}, MSR: ~40 PJ; LSR: ~45 PJ) are already able to cover the current Swiss wood fuel demand for the next 15 years. Considering the "expected" or the more optimistic "maximum" scenarios the strategies fostering stock reductions (MSR, LSR) offer additional potentials on a temporary basis during the energy transition phase. In contrast to the LSR, however, the MSR does not result in a considerable decline in wood fuel from forests expected in 2050 (Table 5). These alternative strategies could result in larger shares of sustainable woody biomass (SPs) compared to the actual annual gross energy consumption of 3.7–6.5% (MSR) and 4.0–7.1% (LSR) in the next 15 years (2020, 2035).

Table 5. Overview of sustainable potentials (SPs) of woody biomass and the different forest management strategies for the three time points (2020, 2035, and 2050). Shown are the minimum and maximum values in [PJ] primary energy. For more details (e.g., different potentials, scenarios and shares), see Supplementary Information S5, Table S12 (TPs); Supplementary Information S5, Table S13 (SPs); and Supplementary Information S6, Table S14 (shares of woody biomass types, TPs and SPs).

Forest Management	2020 [PJ]	2035 [PJ]	2050 [PJ]
CSI [*1]	34.1–59.5	33.6–64.8	29.8–60.3
MSR [*1]	40.1–64.5	42.7–71.8	30.4–59.4
LSR [*1]	45.9–75.1	43.5–77.8	26.5–52.9

[*1] CSI: continuous stock increase (business as usual); MSR: moderate stock reduction; LSR: large stock reduction.

Burg et al. [39] assessed TPs and SPs of future wet biomass (non-woody) in Switzerland using a similar approach. Starting with the current biomass potentials [30], they estimated future availability and variation of resources by taking into account selected drivers and their projected future development. Combining the results of our study with those form [20,30,39] allowed current and future potentials for all biomass types (woody and non-woody) to be presented in one graph (Supplementary Information S6, Figure S1; Supplementary Information S6, Table S14). The share of sustainable woody biomass compared to all sustainable biomass is 42–64% for the next 15 years and declines to 37–57% by 2050. At the same time, all domestic biomass contributes to the Swiss gross annual energy consumption, with sustainable potentials between 7% (72 PJ) and 11% (122 PJ).

4.2. Alternative Forest Wood Management Strategies

Forest wood and the lower wood fuel potentials of trees outside forests are the source of all domestic woody biomass and its subsequent uses. As forest wood contributes the largest share, the amount available for wood fuel use depends largely on the forest management strategy applied. Among the three strategies examined here, MSR tends to result in faster growth by reducing the share of mature forests without hampering forest resources and future wood potentials [20]. Shorter rotation time results in younger forests, with faster growth, and a significant increase in the amount of wood suitable for energy or material use. This also results in smaller amounts of carbon stocked in the above-ground biomass. Increasing wood turnover in forests may lead to significantly higher carbon sequestration rates, larger harvesting amounts, and additional carbon storage in wooden products if subsequent timber industries are able to process additional wood amounts and emission- and energy-intensive materials and fuel are substituted with wooden material. The coupling of wood harvests from forests with the potential from wood residues (Section 2.2.3) implies, however, the domestic processing of surpluses. We assumed that larger wood harvests under a stock reduction scenario (MSR) can be processed by the industry. Evidence for this assumption at least in the short term is given, for example, by the large wood amounts derived during past storm events such as Kyrill (2007) or Burglind (2018). In the subsequent years, cutting amounts in saw mills significantly increased due to salvage logging [52,81], despite the steady decline of processing companies (N = 958, 1991 vs. N = 347, 2017; [52]). Benefits to decarbonization strategies of the Swiss government [82] and the creation of local and sustainable value-added chains are largely dependent on waste wood processing efficiency, wood storage effects (in the case of biogenic carbon accounting), and available cascading options [15].

To reduce the environmental impacts of wood use, Mehr et al. [15] suggest a high-quality wood cascade of wooden beams as a promising recycling path. Other evidence for an implementation of the MSR strategy is that it is expected to be favored by the forestry and timber industry because it leads to improved stability of forest stands against windthrow, pathogen calamities and forest fires, as well as greater diversity of structures and species, and enhanced flexibility in changing tree species. This is particularly seen as a consequence when MSR is combined with shorter rotation times or in response to climate change [20,83]. Today, the degree of salvage loggings in Switzerland due to severe bark beetle calamities (1.4 M m^3 in 2019/2020) [84] points to a more intensive management. Large-scale integration of stock reduction management in Switzerland is difficult due to the small structures of forest enterprises and large number of small private forest owners, as well as a lack of wood demand [85], but this strategy could benefit from Switzerland's large share of public forest area ownership of 71% [73]. Despite its short-term advantages from the view of the potentials alone, the high-intensity management strategy LSR would result in a significant restructuring of forests due to 40% shorter rotation times [20] and accordingly thinner wood compartments, limiting the substitution of more energy-intensive and environmentally harmful materials such as concrete [86]. In the long term (2050), limited possibilities for the substitution of emission- and energy-intensive materials may also lead to less wood fuel from the wood-processing industries (wood residues) and the end-of-life products (waste wood). Therefore, long-term carbon storage in forests and wooden products and added values from woodworking industries are at risk.

Harvesting transportation and chipping costs largely influence the SPs of the different forest managements [20]. We calculated the ecologically sustainable potential (ESP), which represents a "theoretical" potential where the only limitations considered are those imposed by measures to protect biodiversity (in its legal specifications, such as protected areas; [64]) and dead wood (mortality; [45]), including recommendations on harvest losses ([26]; cf. Section 2.2.1) leading to dead wood for flora and fauna and therefore soil nutrients. In this sense, ESP enables the quantification of an upper limit when costs are not a restriction, or revenues increase. The surpluses of ESPs result compared to the corresponding SPs (forest management strategy, scenario, year) were +32% to +75% for CSI, +39% to +156% for MSR and +17% to +200% for LSR (Supplementary Information S1, Table S4).

As a shift in species influences the calorific value and thus the energy content of the wood. Each additional volume unit of broadleaves instead of coniferous wood results in 40% larger wood fuel per unit if values for broadleaves (1.116 t/m^3; 2160 kWh/t) and for conifers (0.758 t/m^3; 2260 kWh/t) are applied (references cf. Section 2.2.5). Today, the proportion of wood fuel consists of 31% broadleaves. They account today for 72% of logs and 62% of chips [73]. Over the whole time period (2020–2050) within the three scenarios, the variations in broadleaves and conifer proportions are small (<±1%). However, between the scenarios, broadleaves have a share on the stocks of 39% (CSI), 40% (MSR) and 43% (LSR) and on increment of 45% (CSI), 46% (MSR) and 47% (LSR).

Trees outside forests and shrubs cut can also contribute to wood fuel potentials. Thus, it is worth mentioning that the main reasons for utilization are safety aspects (e.g., visibility along streets), increased biodiversity, biotope conservation or flood protection [51,67]. As no detailed inventory on the growth of Swiss domestic trees outside forests was available, we assumed values of forest trees to account for growth changes with altitude and had to rely on expert interviews for growth on optimal locations, conducted in [67]. This explains the assumption of the high uncertainty ranges for the minimum and maximum potentials of 20% (TP$_{2020}$) to 60% (TP$_{2050}$) for wood from trees outside forests (Section 2.2.2). The wide ranges may be of little importance as the share of wood from these landscapes is rather small compared to the amounts of other woody biomass types, but they indicate the high degree of uncertainty given that literature and data availability of the main input parameters in areas outside forests (e.g., annual growth) are extremely limited.

4.3. Level of Wood Fuel Consumption

The assortment of wood compartments on wood markets is highly relevant for the quantity of wood fuel provision, its allocation to material purposes, of use and their added value along the process chain [15,20,50,87]. Today, the allocation of assortments (for energy and material use) is dependent on the demand for material products made from wood and the needs and possibilities for energy production also determine the degree of cascade use and circular economy of wood products. While circular economy focuses on how resources can be kept in the system (minimizing primary resources), the cascade concept places the focus on various (hierarchical) end-of-life utilization options. By focusing on wood and other bio-based resources, a cascade concept could be considered to form a connection between circular economy and the (wood or) bio-economy, fostering an inclusive, circular bio-economic vision in the future [88]. Therefore, the discussion on applying a less wood energy-oriented (EO−) or wood energy-oriented (EO+) market situation, including the subsequent woodworking industries, may offer integration strategies for wood fuel use in cascade and circular concepts.

The EO− market situation assorts larger shares of wood compartments (roundwood) for material processing. This increased use of wood for materials directly after harvests creates additional wood residues from processing but also allows for the reuse and recycling of wood residues and wooden products [15]. The basis for reuse and recycling is given by physical properties of wooden products, such as particle size or the presence of chemicals [89]. In an EO− market, this is 20–30% of the thicker R4–R6 (0.40–>0.60 m) compartments and 15–20% of the thinner R1–R3 (0.10–0.39 m) compartments in the case of conifers and for 30–40% of all roundwood (R2–R6, 0.20–>0.60 m) compartments for broadleaves (Supplementary Information S1, Table S5). No quantities for material use are considered for bark, brushwood and the thinnest class of broadleaves (R1, 0.1–0.19 m) in either market situation [20]. This may overestimate the sustainable wood fuel potentials for energy from forests because small wood parts can currently also be used for material processing [90]. For the other compartments, the EO− market requires (as also surpluses from forest management) the necessary processing capacity of sawmills and the corresponding demand for sawn timber of the subsequent wood-processing companies. In Switzerland today, some processing industries are absent to account for complete regional cascades [13]. In particular, many material-processing companies have closed (e.g., PAVATEX) or moved abroad due to excessively high operating costs, e.g., due to the strong Swiss Franc [91]. These industries are considered important for implementing full concepts and combining circular economy with cascade options including final energy use. However, the price of wood energy has been high for many years and exceeds the price of industrial wood [85], and it is also a consequence of the huge amounts of imported (finished) wood products, as production costs abroad are lower. This may lead to even lower prices for domestic wood materials and construction wood and promote energy use of also thicker wood compartments directly after harvests. In particular, imports are concentrated on the amounts of finished or semi-finished wooden products [92] and therefore reduce the need for domestic wood for material processing. As a result, domestic wood resources are sometimes used too early in the process chain to optimize wood use for energy (e.g., in terms of full carbon storage, added value or substitution). This means wood suitable for material purposes is used for energy instead. Furthermore, large transportation distances hinder sustainable and climate-smart use of wood trading. By exporting the raw timber, coupled wood residues and future wastes occurring within the next processing steps (and their added value) are also lost, as residues and wastes occur abroad in the subsequent timber processing steps [51]. Finally, the importing of semi-finished products, sawn timber, plywood, chipboard, fiberboard, construction and packaging materials, wooden products and furniture (e.g., IKEA) leads—after further processing or at the end of their life cycle—to additional potential wood fuel quantities from abroad. By assessing waste wood directly at the facility sites, our estimates also take into consideration these imported and exported quantities.

From an environmental point of view, waste wood processing efficiencies and available cascade options are decisive for the total environmental impacts [15], as they largely depend on the production structure, environmental circumstances, local policies, regulations and economic resources [37]. It is particularly relevant which other products are substituted during the cascade and how completely and efficiently wood can be used for energy production in the end. In terms of greenhouse gas emissions, wood use is beneficial in most applications. Benefits are particularly high when replacing heat from oil or gas, substituting for energy-intensive products such as primary metals, plastic and concrete in constructions and furniture [92], or functioning in difficult-to-decarbonize sectors such as aviation, heavy transportation (e.g., trucks, forwarders, agricultural machines), and manufacturing [6]. Therefore, it is important to consider what for the substituted, often non-renewable (e.g., fossil) products are used for and whether they are able to support the production and supply security of high value products (e.g., medicines, paints, notebooks, car fittings) and therefore potentially have valuable social benefits.

In implementing the energy use of wood, special requirements regarding the preparation and recycling of already used and treated wood (waste wood) have to be considered [53,93–95]. Furthermore, the orientation towards an EO– market would support continuously and significantly larger resource quantities for one or more processing steps. In this way, most of the resource stays available as a storable, regional (depending on the distribution of energy plants and locations of SPs) energy carrier in future, simultaneously increasing the carbon storage in wooden products [96]. However, the use of waste wood and wood residues for energy takes place within a certain timeframe [97]. Therefore, the desired convergence towards a cascade and more circular economy in Switzerland [13] and Europe (e.g., [89,98]) could temporarily lower wood fuel potentials from residues because of better integration of processed and already used wood for product processing in (innovative) cascades, closed loops and recycling, innovative collaborations and products [88,99]. Later on, when wood residues are used more efficiently for materials, a shift in wood fuel from wood residues towards wood fuel potentials from waste wood can be expected. Wood from trees outside forests are mainly thinner compartments from shrubs, trees and hedges, especially along water resources, roads, paths, railway lines and building yards, orchards and vineyards or bush and shrub vegetation. Therefore, the potential for material use is particularly low (e.g., 7% according to [67]).

4.4. Other Key Drivers and Critical Remarks

With our analyses, we aimed to quantify woody biomass potentials and their development with regard to their original source and realistic ranges. However, the different key drivers and the determination of future sustainable potentials were combined with different restrictions relevant for the individual biomass types. This makes it difficult to evaluate the effects of the individual influences on the potentials. Therefore, we concentrated on forest management strategies with different harvesting costs (Sections 4.1 and 4.2) and material uses with different wood markets (circular or cascading concepts) (Section 4.3). Nonetheless, other drivers and restrictions have an influence. We rely mainly on various literature values (Sections 2.2.1–2.2.4), but their qualitative validation provides confidence in their accuracy.

Climate change influences the growth of forests and timber positively (e.g., temperature) and negatively (e.g., drought) via biotic and abiotic factors. Furthermore, nitrogen deposition, among others, affect the primary growing conditions of forests well below a classical rotation period [100–103]. In our simulations, we expected site characteristics promoting wood fuel from forests and trees outside forests to be favored in the future due to climate change. Overall, we expect climate change to lead to an additional net growth of 11% in 2035 and 10% in 2050 (cf. Section 2.2.1) [49,62]. This adds additional wood fuel biomass to TPs of 10.4–15.0 PJ in 2035 and 9.4–13.6 PJ in 2050. The lowest additional amounts are based on the forest management of CSI and the largest on LSR. However, the effect of changing site conditions on the potentials of woody biomass from forests may be even greater than the total woody biomass coming from trees outside forests in the long term.

The change in forest area size of +3% (2035) and +6% (2050) is more pronounced in coline and mountainous regions (>12%, [56]). This is particularly true where, e.g., forest limits move upwards due to climate change [104] or where land-use change occurs—for example, the abandoned Alps are less and less cultivated by summering cattle mainly due to more fodder availability on the home farm [105]. Forests will have to adapt not only to changes in mean climate variables but also with a likely increased risk of extreme weather events, such as prolonged drought, storms and floods (e.g., [106]). Especially in northern and western Europe, the increasing atmospheric CO_2 content and warmer temperature are expected to result in positive effects on forest growth and wood production, at least in the short–medium term. On contrary, more frequent droughts and disturbance risks will cause adverse effects [107]. For example, ref. [108] shows high vulnerability particularly in the second half of the twenty-first century in the eastern Alps. They indicate a strong decline in productivity, timber and carbon stocks, biodiversity, disturbances, a tree species' position in fundamental niche space, silvicultural flexibility and cost intensity. Most negatively affected were sites on calcareous bedrock, whereas assessment units at higher altitudes responded predominately positively to an increase in temperature. Such negative impacts may outweigh positive trends [107]. Therefore, a change of disturbance events (e.g., storms or droughts) may consequently change increment and mortality and may outweigh other key drivers of the simulation. Regions with longer drought periods and more frequent and violent thunderstorms, as well as those where reforestations occur for economic reasons, can also be expected to see larger changes in forest area size. Thus, increased productivity of the sward and longer summer farming periods, as well as subsidies, aim to maintaining the current status and increase the ecological quality of pastures [105]. In contrast, in the Swiss Central Plateau, where forest areas remain stable [17], the pressure on forests is constantly increasing, especially as the growth in population is accompanied by a growing demand for settlements and infrastructures. Since the forests in Switzerland are protected by law [109], the changes outside forests mainly occur at the expense of barely cultivated agricultural land. This supports our assumption of a continued increase in stocked areas of trees outside forests due to the extension of settlement areas with yards or greenings (stocked areas). We based our assumption on past development of trees outside forests (Section 2.2.2), which corresponds to a +0.46% stocked area annually. However, as building land reserves are limited and the desired densification of already zoned areas are intended by law [110], the assumption of a stagnation in non-forest landscape area growth in 2035 seems realistic as, for example, no further stocked surroundings of building are expected.

Furthermore, the approach considering wood fuel potentials according to their original sources has the advantage that the volumes of wood residues are determined in accordance with changes in forest management or energy markets, while simultaneously including imported end products of wood occurring with a time lag as waste wood. However, this also means that volumes of wood residues within this study are not directly related to the woodworking and processing companies where they de facto occur. Therefore, wood fuel from wood residues contribute only a very small share of SPs (4–14%) compared to TPs. This is because TPs of wood residues are calculated according to their original forest sources and therefore are related to TPs of harvesting. Related wood residues only occur if wood working companies are able to process these huge amounts. In contrast, SPs of wood fuel from wood residues are calculated with SPs of forest harvesting (Supplementary Information S3, Table S10). They account for restrictions of protected areas, harvest losses, assortments (wood market: EO−) and costs. Additionally, parts of the processed wood (~45%, cf. Figure 3, recycling vs. energy) are used for recycling and occur after several processing steps for energy or as waste wood at the end of the product's life cycle. On the one hand, the modeled wood market EO− simulates a shift towards the more desired, cascade-oriented wood market (cf. Section 4.3). On the other hand, wood residues are kept in the system with circular economy or cascade use and result in at least a temporary shift of wood fuel from wood residues towards waste wood (Section 4.3). As a consequence of this shift, increased wood fuel amounts from waste wood result (2020, 12.9 to 15.8 PJ; 2050, 15.3 to 20.9 PJ, Supplementary Information S4, Table S11). Furthermore, the driver population growth assumes equal wood consumption per person during the coming years to reach higher SPs and TPs in 2035 and 2050 by approximately 20% (cf. Section 2.2.4).

5. Conclusions

Our analysis indicates that wood fuel potentials can be expected to be stable enough to provide sustainable potentials (SPs) within the next 15 years. This is also the case if the original sources are combined with their subsequent woodworking industries, as wood fuel amounts from a minimum of ~34 PJ/a to a temporary maximum of ~78 PJ/a are available. The moderate stock reduction (MSR) forest management alternative leads to larger amounts of wood resources, especially in the short term, without causing future resource shortages of wood fuel and wood for materials compared to the business as usual (CSI) strategy. Simultaneously, the large stock reduction (LSR) offers higher additional potentials in the short term but will likely lead to smaller amounts available for wood fuel in 2050. Additionally, thinner compartments are expected with LSR, leading to a potential risk for the material wood markets. The shorter rotation times of LSR and MSR may increase flexibility in forest management and could provide a more stable basis for dealing with climate change and extremes while increasing the potential for the energy transition. In particular, with the current forced uses of wood in Switzerland due to drought, bark beetle infestation and windthrow as a consequence of climate change and extreme weather events, the actual harvests come very close to the quantitative estimation obtained with the MSR strategy.

When costs are not a restriction (e.g., due to subsidies and more efficient technologies and processes) or revenues increase significantly (e.g., higher prices for energy), the resulting ecologically sustainable potential (ESP) yields large surpluses of approximately one-fifth up to three-fold more (LSR, 2035, scenario maximum; cf. Section 4.2) compared to the corresponding (scenario and point in time) sustainable potential (SP) of the business as usual strategy (CSI). If the wood market is less energy oriented (EO−), circular economy and cascade uses of wood can be supported, as more wood is assorted towards material uses first but is still available for wood fuel with a time lag and a shift in wood fuel form forests towards wastes occurring during wood processing (as wood residues) or ideally at the end of the product's life cycle (as waste wood). For implementation, a more circular economy like this requires the necessary processing capacities of the subsequent timber industries, as well as the energy plants for the wood fuel conversion to heat, electricity, liquid fuel or wood gas. The needed infrastructures could be better planned when wood fuel potentials' development is combined with the desired cascade use and circular economy at a high spatial resolution, enabling more efficient and effective transport of resources from the harvesting locations to the plants. In terms of carbon storage, living biomass as well as storage in subsequent timber products is considered of key importance, particularly if emission- and energy-intensive products are substituted. Shorter rotation periods promoting regrowth and carbon storage in young plants can therefore lead to an increasing carbon stock.

Supplementary Materials: Supplementary Materials are available online at http://www.mdpi.com/2071-1050/12/22/9749/s1.

Author Contributions: Conceptualization, M.E., V.B., O.T. and J.S.; data curation, M.E., O.T. and G.S.; formal analysis, M.E., V.B. and O.T.; funding acquisition, O.T.; investigation, M.E. and O.T.; methodology, M.E., V.B. and O.T.; project administration, O.T.; supervision, O.T. and M.F.; validation, M.E., V.B., L.B., O.T., M.F. and J.S.; visualization, M.E., V.B., L.B., O.T., M.F. and J.S.; writing—original draft, M.E., O.T. and J.S.; writing—review and editing, M.E., V.B., L.B., O.T., M.F., G.S. and J.S. All authors have read and agreed to the published version of the manuscript.

Funding: This research project is part of the Swiss Competence Center for Energy Research—Biomass for Swiss Energy Future (SCCER BIOSWEET), which is financially supported by the Innosuisse—Swiss Innovation Agency.

Acknowledgments: We thank Melissa Dawes and Colm Kelly for supporting the writing process, with English editing and helpful ideas. Furthermore, we thank Gillianne Bowman for her valuable support with figures and illustrations.

Conflicts of Interest: The authors declare no conflict of interest. The funders had no role in the design of the study; in the collection, analyses, or interpretation of data; in the writing of the manuscript, or in the decision to publish the results.

References

1. George, G.; Schillebeeckx, S.J.D.; Liak, T.L. The Management of Natural Resources: An Overview and Research Agenda. *Acad. Manag. J.* **2015**, *58*, 1595–1613.
2. Kirchner, A.; Bredow, F.; Grebel, T.; Hofer, P.; Kemmler, A.; Ley, A.; Piégsa, A.; Schütz, N.; Strassburg SStruwe, J. *Die Energieperspektiven für die Schweiz bis 2050, Energienachfrage und Elektrizitätsangebot in der Schweiz 2000–2050*; Prognos AG: Basel, Switzerland, 2012; p. 842.
3. Long, H.; Li, X.; Wang, H.; Jia, J. Biomass resources and their bioenergy potential estimation: A review. *Renew. Sustain. Energy Rev.* **2013**, *26*, 344–352.
4. Li, Y.; Rezgui, Y.; Zhu, H. District heating and cooling optimization and enhancement—Towards integration of renewables, storage and smart grid. *Renew. Sustain. Energy Rev.* **2017**, *72*, 281–294.
5. Lauri, P.; Havlík, P.; Kindermann, G.; Forsell, N.; Böttcher, H.; Obersteiner, M. Woody biomass energy potential in 2050. *Energy Policy* **2013**, *66*, 19–31.
6. Reid, W.V.; Ali, M.K.; Field, C.B. The future of bioenergy. *Glob. Chang. Biol.* **2020**, *26*, 274–286.
7. European Academies Science Advisory Council. *Multi-Functionality and Sustainability in the European Union's Forests*; German National Academy of Sciences: Halle, Germany, 2017; p. 51.
8. Sigsgaard, T.; Forsberg, B.; Annesi-Maesano, I.; Blomberg, A.; Bølling, A.; Boman, C.; Bønløkke, J.; Brauer, M.; Bruce, N.; Héroux, M.-E.; et al. Health impacts of anthropogenic biomass burning in the developed world. *Eur. Respir. J.* **2015**, *46*, 1577–1588.
9. Barreiro, S.; Schelhaas, M.-J.; Kändler, G.; Antón-Fernández, C.; Colin, A.; Bontemps, J.-D.; Alberdi, I.; Condés, S.; Dumitru, M.; Ferezliev, A.; et al. Overview of methods and tools for evaluating future woody biomass availability in European countries. *Ann. For. Sci.* **2016**, *73*, 823–837.
10. Schuldt, B.; Buras, A.; Arend, M.; Vitasse, Y.; Beierkuhnlein, C.; Damm, A.; Gharun, M.; Grams, T.E.; Hauck, M.; Hajek, P.; et al. A first assessment of the impact of the extreme 2018 summer drought on Central European forests. *Basic Appl. Ecol.* **2020**, *45*, 86–103.
11. Bowditch, E.; Santopuoli, G.; Binder, F.; Del Río, M.; La Porta, N.; Kluvankova, T.; Lesinski, J.; Motta, R.; Pach, M.; Panzacchi, P.; et al. What is Climate-Smart Forestry? A definition from a multinational collaborative process focused on mountain regions of Europe. *Ecosyst. Serv.* **2020**, *43*, 101113.
12. Schweier, J.; Magagnotti, N.; Labelle, E.R.; Athanassiadis, D. Sustainability Impact Assessment of Forest Operations: A Review. *Curr. For. Rep.* **2019**, *5*, 101–113.
13. Federal Office for the Environment (FOEN); Swiss Federal Office of Energy (SFOE); State Secretariat for Economic Affairs (SECO). *(Eds.) Ressourcenpolitik Holz. Strategie, Ziele und Aktionsplan Holz*; Federal Office of the Environment (FOEN)/Federal Office of Energy (SFOE)/State Secretariat for Economic Affairs (SECO): Bern, Switzerland, 2018; p. 62.
14. Federal Office for the Environment (FOEN); Swiss Federal Office of Energy (SFOE); State Secretariat for Economic Affairs (SECO). *(Eds.) Ressourcenpolitik Holz. Strategie, Ziele und Aktionsplan Holz*; Federal Office of the Environment (FOEN)/Federal Office of Energy (SFOE)/State Secretariat for Economic Affairs (SECO): Bern, Switzerland, 2014; p. 36.
15. Mehr, J.; Vadenbo, C.; Steubing, B.; Hellweg, S. Environmentally optimal wood use in Switzerland—Investigating the relevance of material cascades. *Resour. Conserv. Recycl.* **2018**, *131*, 181–191.
16. Farrell, E.P.; Führer, E.; Ryan, D.; Andersson, F.; Hüttl, R.; Piussi, P. European forest ecosystems: Building the future on the legacy of the past. *For. Ecol. Manag.* **2000**, *132*, 5–20.
17. Brändli, U.B.; Abegg, M.; Allgaier, L. *(Red.), Swiss National Forest Inventory, Results of the Fourth Survey 2009–2017*; Swiss Federal Institute for Forest, Snow and Landscape Research WSL: Birmensdorf, Switzerland; Federal Office of the Environment (FOEN): Bern, Switzerland, 2020; p. 341.
18. Schulze, E.-D.; Sierra, C.A.; Egenolf, V.; Woerdehoff, R.; Irslinger, R.; Baldamus, C.; Stupak, I.; Spellmann, H. The climate change mitigation effect of bioenergy from sustainably managed forests in Central Europe. *GCB Bioenergy* **2020**, *12*, 186–197.

19. Zhang, F.; Johnson, D.M.; Wang, J.; Liu, S.; Zhang, S. Measuring the Regional Availability of Forest Biomass for Biofuels and the Potential of GHG Reduction. *Energies* **2018**, *11*, 198.

20. Thees, O.; Erni, M.; Lemm, R.; Stadelmann, G.; Zenner, E.K. Future potentials of sustainable wood fuel from forests in Switzerland. *Biomass Bioenergy* **2020**, *141*, 105647.

21. Wörsdörfer, M. *The Future of Petrochemicals, Director of Sustainability, Technology and Outlooks*; Brussels-EU Refining Forum: Brussels, Belgium, 2018.

22. Hu, H.; Wu, M. Heavy oil-derived carbon for energy storage applications. *J. Mater. Chem. A* **2020**, *8*, 7066–7082.

23. Searle, S.; Malins, C. A reassessment of global bioenergy potential in 2050. *GCB Bioenergy* **2015**, *7*, 328–336.

24. Bončina, A.; Simončič, T.; Rosset, C. Assessment of the concept of forest functions in Central European forestry. *Environ. Sci. Policy* **2019**, *99*, 123–135.

25. Trutnevyte, E.; Guivarch, C.; Lempert, R.; Strachan, N. Reinvigorating the scenario technique to expand uncertainty consideration. *Clim. Chang.* **2016**, *135*, 373–379.

26. Hepperle, F. *Prognosemodell zur Abschätzung des Regionalen Waldenergieholzpotenzials auf der Grundlage Forstlicher Inventur- und Planungsdaten unter Berücksichtigung Ökologischer, Technischer und Wirtschaftlicher Nutzungseinschränkungen*; Fakultät für Forst und Umweltwissenschaften der Albert-Ludwigs-Universität Freiburg im Breisgau: Freiburg im Breisgau, 2010; p. 165.

27. Baier, U.; Baum, S. *Biogene Güterflüsse der Schweiz 2006, Massen- und Energieflüsse*; Bundesamt für Umwelt (BAFU): Bern, Switzerland, 2006; p. 114.

28. Balat, M. Use of biomass sources for energy in Turkey and a view to biomass potential. *Biomass Bioenergy* **2005**, *29*, 32–41.

29. Steubing, B.; Zah, R.; Waeger, P.; Ludwig, C. Bioenergy in Switzerland: Assessing the domestic sustainable biomass potential. *Renew. Sustain. Energy Rev.* **2010**, *14*, 2256–2265.

30. Burg, V.; Bowman, G.; Erni, M.; Lemm, R.; Thees, O. Analyzing the potential of domestic biomass resources for the energy transition in Switzerland. *Biomass Bioenergy* **2018**, *111*, 60–69.

31. Thrän, D.; Seidenberger, T.; Zeddies, J.; Offermann, R. Global biomass potentials—Resources, drivers and scenario results. *Energy Sustain. Dev.* **2010**, *14*, 200–205.

32. Verkerk, P.J.; Fitzgerald, J.B.; Datta, P.; Dees, M.; Hengeveld, G.M.; Lindner, M.; Zudin, S. Spatial distribution of the potential forest biomass availability in Europe. *For. Ecosyst.* **2019**, *6*, 1–11.

33. Vavrova, K.; Knapek, J.; Weger, J. Short-term Boosting of Biomass Energy Sources—Determination of Biomass Potential for Prevention of Regional Crisis Situations. *Renew. Sustain. Energy Rev.* **2017**, *67*, 426–436.

34. Gustavsson, L.; Sathre, R. Variability in energy and carbon dioxide balances of wood and concrete building materials. *Build. Environ.* **2006**, *41*, 940–951.

35. Hagemann, N.; Gawel, E.; Purkus, A.; Pannicke, N.; Hauck, J. Possible Futures towards a Wood-Based Bioeconomy: A Scenario Analysis for Germany. *Sustainability* **2016**, *8*, 98.

36. Vadenbo, C.; Tonini, D.; Burg, V.; Astrup, T.F.; Thees, O.; Hellweg, S. Environmental optimization of biomass use for energy under alternative future energy scenarios for Switzerland. *Biomass Bioenergy* **2018**, *119*, 462–472.

37. Heräjärvi, H.; Kunttu, J.; Hurmekoski, E.; Hujala, T. Outlook for modified wood use and regulations in circular economy. *Holzforschung* **2019**, *2019*, 10.

38. Aylott, M.J.; Casella, E.; Tubby, I.; Street, N.R.; Smith, P.; Taylor, G. Yield and spatial supply of bioenergy poplar and willow short-rotation coppice in the UK. *New Phytol.* **2008**, *178*, 358–370.

39. Burg, V.; Bowman, G.; Hellweg, S.; Thees, O. Long-Term Wet Bioenergy Resources in Switzerland: Drivers and Projections until 2050. *Energies* **2019**, *12*, 3585.

40. Werner, C.; Haas, E.; Grote, R.; Gauder, M.; Graeff-Hönninger, S.; Claupein, W.; Butterbach-Bahl, K.; Grote, R. Biomass production potential fromPopulusshort rotation systems in Romania. *GCB Bioenergy* **2012**, *4*, 642–653.

41. Aust, C.; Schweier, J.; Brodbeck, F.; Sauter, U.H.; Becker, G.; Schnitzler, J.-P. Land availability and potential biomass production with poplar and willow short rotation coppices in Germany. *GCB Bioenergy* **2014**, *6*, 521–533.

42. Van Meerbeek, K.; Muys, B.; Hermy, M. Lignocellulosic biomass for bioenergy beyond intensive cropland and forests. *Renew. Sustain. Energy Rev.* **2019**, *102*, 139–149.

43. Brosowski, A.; Thrän, D.; Mantau, U.; Mahrod, B.; Erdmann, G.; Adlera, P.; Stinner, W.; Reinholde, G.; Heringe, T.; Blankec, C. A Review of Biomass Potential and Current Utilisation—Status quo for 93 Biogenic Wastes and Residues in Germany. *Biomass Bioenergy* **2016**, *95*, 257–272.

44. Mantau, U.; Saal, U.; Prins, K.; Streierer, F.; Lindner, M.; Verkerk, H.; Eggers, J.; Leek, N.; Oldenburger, J. *EUwood—Real Potential for Changes in Growth and Use of EU Forests 2010, Final Report*; EFI: Hamburg, Germany, 2010; p. 160.

45. Stadelmann, G.; Temperli, C.; Rohner, B.; Didion, M.; Herold, A.; Rösler, E.; Thürig, E. Presenting MASSIMO: A Management Scenario Simulation Model to Project Growth, Harvests and Carbon Dynamics of Swiss Forests. *Forests* **2019**, *10*, 94.

46. Brändli, U.-B. (Ed.) *Ergebnisse der Dritten Erhebung 2004–2006*; Schweizerisches Landesforstinventar; Eidg. Forschungsanstalt für Wald, WSL: Birmensdorf, Switzerland; Bundesamt für Umwelt (BAFU): Bern, Switzerland, 2010; p. 316.

47. Frutig, F.; Thees, O.; Lemm, R.; Kostadinov, F. *Holzernteproduktivitätsmodelle HeProMo-Konzeption, Realisierung, Nutzung und Weiterentwicklung, Grundlagen, Methoden und Instrumente*; Eidg. Forschungsanstalt WSL: Birmensdorf, Switzerland, 2009; pp. 441–466.

48. Holm, S.; Frutig, F.; Lemm, R.; Thees, O.; Schweier, J. HeProMo: A Decision Support Tool to Estimate Wood Harvesting Productivities. *PLoS ONE* **2020**. Submitted.

49. Federal Office for the Environment (FOEN). *Faktenblatt zur Strategie des Bundesrates «Anpassung an den Klimawandel in der Schweiz»*; Anpassung an den Klimawandel; Sektor Waldwirtschaft: Bern, Switzerland, 2013.

50. Thees, O.; Kaufmann, E.; Lemm, R.; Bürgi, A. Energieholzpotenziale im Schweizer Wald. *Schweiz. Z. Forstwes.* **2013**, *164*, 351–364.

51. Thees, O.; Burg, V.; Erni, M.; Bowman, G.; Lemm, R. *Biomassenpotenziale der Schweiz für die Energetische Nutzung, Ergebnisse des Schweizerischen Energiekompetenzzentrums SCCER BIOSWEET*; Eidg. Forschungsanstalt für Wald Schnee und Landschaft (WSL): Birmensdorf, Switzerland, 2017; p. 299.

52. Federal Statistical Office (FSO). *Entwicklung des Rundhzolzeinschittes in den Sägereien Nach Kantonen*; FSO: Neuchâtel, Switzerland, 2018.

53. Der Schweizerische Bundesrat, Verordnung über den Verkehr mit Abfällen (VeVA). 2005. (Fassung Juli 2016). Available online: https://docplayer.org/21490070-Verordnung-ueber-den-verkehr-mit-abfaellen.html (accessed on 18 November 2020).

54. Stadelmann, G.; Herold, A.; Didion, M.; Vidondo, B.; Gomez, A.; Thürig, E. Holzerntepotenzial im Schweizer Wald: Simulation von Bewirtschaftungsszenarien. *Schweiz. Z. Forstwes.* **2016**, *167*, 152–161.

55. Temperli, C.; Stadelmann, G.; Thürig, E.; Brang, P. Timber Mobilisation and Habitat Tree Retention in Low-Elevation mixed Forests in Switzerland: An Inventory-based Scenario Analysis of Opportunities and Constraints. *Eur. J. For. Res.* **2017**, *136*, 711–725.

56. Ginzler, C.; Brändli, U.-B.; Hägeli, M. Waldflächenentwicklung der letzten 120 Jahre in der Schweiz. *Schweiz. Z. Forstwes.* **2011**, *162*, 337–343.

57. Bircher, N.; Cailleret, M.; Huber, M.; Bugmann, H. Empfindlichkeit typischer Schweizer Waldbestände auf den Klimawandel. *Schweiz. Z. Forstwes.* **2015**, *166*, 408–419.

58. Mina, M.; Bugmann, H.; Cordonnier, T.; Irauschek, F.; Klopčič, M.; Pardos, M.; Cailleret, M. Future ecosystem services from European mountain forests under climate change. *J. Appl. Ecol.* **2017**, *54*, 389–401.

59. Rohner, B.; Kumar, S.; Liechti, K.; Gessler, A.; Ferretti, M. Tree Vitality Indicators Revealded a Reapid Response of Beech Forests to the 2018 Drought. *Ecol. Indic.* **2021**, *120*, 106903.

60. Rohner, B.; Waldner, P.; Lischke, H.; Ferretti, M.; Thürig, E. Predicting individual-tree growth of central European tree species as a function of site, stand, management, nutrient, and climate effects. *Eur. J. For. Res.* **2018**, *137*, 29–44.

61. Trotsiuk, V.; Hartig, F.; Cailleret, M.; Babst, F.; Forrester, D.I.; Baltensweiler, A.; Buchmann, N.; Bugmann, H.; Gessler, A.; Gharun, M.; et al. Assessing the response of forest productivity to climate extremes in Switzerland using model–data fusion. *Glob. Chang. Biol.* **2020**, *26*, 2463–2476.

62. Nabuurs, G.-J.; Pussinen, A.; Karjalainen, T.; Erhard, M.; Kramer, K. Stemwood volume increment changes in European forests due to climate change—A simulation study with the EFISCEN model. *Glob. Chang. Biol.* **2002**, *8*, 304–316.

63. Abegg, M.; Brändli, U.-B.; Cioldi, F.; Fischer, C.; Herold-Bonardi, A.; Huber, M.; Keller, M.; Meile, R.; Rösler, E.; Speich, S.; et al. *Schweizerisches Landesforstinventar—Ergebnistabelle Nr. 173534: Jährlicher Nettozuwachs*; Eidg. Forschungsanstalt WSL: Birmensdorf, Switzerland, 2014.

64. Losey, S.; Wehrli, A. *Schutzwald in der Schweiz. Vom Projekt SilvaProtect-CH zum harmonisierten Schutzwald*; Bundesamt für Umwelt (BAFU): Bern, Switzerland, 2013; p. 29.

65. Federal Office for the Environment (FOEN). *Waldpolitik 2020, Visionen, Ziele, Massnahmen für Eine Nachhaltige Bewirtschaftung des Schweizer Waldes*; FOEN: Bern, Switzerland, 2013; p. 66.

66. Federal Statistical Office (FSO). *Areal Statistics Switzerland, Standard Nomenclature, Data Set, Hectare Level*; FSO: Neuchâtel, Switzerland, 2009.

67. Federal Office for the Environment (FOEN); Swiss Federal Office of Energy (SFOE). *Energieholzpotenziale Ausserhalb des Waldes*; Federal Office for the Environment (FOEN)/Swiss Federal Office of Energy (SFOE): Bern, Switzerland, 2009; p. 82.

68. Federal Statistical Office (FSO). *Digital Elevation Model (100 m)*; GEOSTAT: Neuchâtel, Switzerland, 1979.

69. Swisstopo Federal Office of Topography. *Digital Land Model of Switzerland, Vector 25*; Swisstopo Federal Office of Topography: Bern, Switzerland, 2007.

70. Federal Office for the Environment (FOEN). *Jahrbuch Wald und Holz*; Federal Office for the Environment (FOEN): Bern, Switzerland, 2014; p. 162.

71. Erni, M.; Thees, O.; Lemm, R. *Altholzpotenziale der Schweiz für die Energetische Nutzung, Ergebnisse einer Vollerhebung*; WSL Ber. 52; Eidg. Forschungsanstalt WSL: Birmensdorf, Switzerland, 2017; p. 68.

72. Mantau, U.; Weimar, H.; Kloock, T. *Standorte der Holzwirtschaft, Holzrohstoffmonitoring, Altholz im Entsorgungsmarkt, Aufkommens- und Vertriebsstruktur 2010*; Universität Hamburg: Hamburg, Germany, 2012; p. 30.

73. Federal Office for the Environment (FOEN). *Jahrbuch Wald und Holz*; Federal Office for the Environment (FOEN): Bern, Switzerland, 2018.

74. Kohli, R.; Bläuer, A.; Perrenoud, S.; Babel, J. *Szenarien zur Bevölkerungsentwicklung der Schweiz 2015–2045*; Bundesamt für Statistik (BFS): Neuchâtel, Switzerland, 2015; p. 82.

75. Taverna, R.; Meister, R.; Hächler, K. *Abschätzung des Altholzaufkommens und des CO$_2$-Effektes aus Seiner Energetischen Verwertung, Schlussbericht, Bundesamt für Umwelt (BAFU)*; Departement für Umwelt, Verkehr, Energie und Kommunikation (UVEK): Bern, Switzerland, 2010; p. 63.

76. Lehner, L.; Kinnunen, U.; Weidner, J.; Lehner, J.; Pauli, B.; Menk, J. *Branchenanalyse, Analyse und Synthese der Wertschöpfungskette (WSK) Wald und Holz in der Schweiz, Bern, Bundesamt für Umwelt (BAFU)*; Eidg. Departements für Umwelt, Verkehr, Energie und Kommunikation (UVEK): Bern, Switzerland, 2014; p. 345.

77. Hahn, J.; Schardt, M.; Schulmeyer, F.; Mergler, F. *Der Energieinhalt von Holz*; Bayerische Landesanstalt für Wald und Forstwirtschaft (LWF): Freising, Germany, 2014; p. 4.

78. Swiss Federal Office of Energy (SFOE). *Masse, Einheiten, Zahlen: Umrechnungsfaktoren, Masseinheiten und Energieinhalte*; SFOE: Bern, Switzerland, 2006; p. 1.

79. Federal Statistical Office (FSO). *Restholzverwertung in den Sägereien nach Kantonen 2012, Tab 7.3.5.3*; FSO: Neuchâtel, Switzerland, 2013.

80. Federal Office of Energy (SFOE). *Überblick über den Energieverbrauch der Schweiz im Jahr 2019, Auszug aus der Schweizerischen Gesamtenergiestatistik 2019*; SFOE: Bern, Switzerland, 2020; p. 9.

81. MeteoSchweiz. *Der Wintersturm Burglind/Eleanor in der Schweiz, Fachbericht*; MeteoSchweiz: Zurich, Switzerland, 2018; p. 35.

82. Marcucci, A.; Zhang, L. Growth impacts of Swiss steering-based climate policies. *Swiss J. Econ. Stat.* **2019**, *155*, 1–13.

83. Yousefpour, R.; Temperli, C.; Jacobsen, J.B.; Thorsen, B.J.; Meilby, H.; Lexer, M.J.; Lindner, M.; Bugmann, H.; Borges, J.G.; Palma, J.H.; et al. A framework for modeling adaptive forest management and decision making under climate change. *Ecol. Soc.* **2017**, *22*, 40.

84. Stroheker, S.; Forster, B.; Queloz, V. *Zweithöchster je Registrierter Buchdruckerbefall (Ips Typographus) in der Schweiz*; Waldschutz Aktuell; Eidg. Forschungsanstalt WSL: Birmensdorf, Switzerland, 2020; p. 2.

85. Federal Office of the Environment (FOEN). *2018 Market Statement for Switzerland, Developments in Forest Product Markets*; FOEN: Bern, Switzerland, 2018; p. 31.

86. Ede, A.N.; Adebayo, S.O.; Ugwu, E.I.; Emenike, C. Life Cycle Assessment of Environmental Impacts of using Concrete or Timber to Construct a Duplex Residential Building. *IOSR J. Mech. Civ. Eng.* **2014**, *11*, 62–72.

87. Abasian, F.; Rönnqvist, M.; Ouhimmou, M. Forest bioenergy network design under market uncertainty. *Energy* **2019**, *188*, 116038.

88. Mair-Bauernfeind, C.; Stern, T. Cascading Utilization of Wood: A Matter of Circular Economy? *Curr. For. Rep.* **2017**, *3*, 281–295.

89. Jarre, M.; Petit-Boix, A.; Priefer, C.; Meyer, R.; Leipold, S. Transforming the Bio-based Sector Towards a Circular Economy—What can we learn from Wood Cascading? *For. Policy Econ.* **2020**, *110*, 1–12.

90. Shahab, R.L.; Brethauer, S.; Davey, M.P.; Smith, A.G.; Vignolini, S.; Luterbacher, J.S.; Studer, M. A heterogeneous microbial consortium producing short-chain fatty acids from lignocellulose. *Science* **2020**, *369*, 9.

91. Holzindustrie Schweiz (HIS). *Annual Report of the European Sawmill Industry 2018/2019, EOS ANNUAL REPORT, Brumunddal, Norway*; HIS: Bern, Switzerland, 2019; p. 226.

92. Suter, F.; Steubing, B.; Hellweg, S. Life Cycle Impacts and Benefits of Wood along the Value Chain: The Case of Switzerland. *J. Ind. Ecol.* **2016**, *21*, 874–886.

93. Röder, M.; Thornley, P. Waste wood as bioenergy feedstock. Climate change impacts and related emission uncertainties from waste wood based energy systems in the UK. *Waste Manag.* **2018**, *74*, 241–252.

94. Der Schweizerische Bundesrat. *Luftreinhalte-Verordnung (LRV)*; Der Schweizerische Bundesrat: Bern, Switzerland, 2016; p. 98.

95. Der Shweizerische Bundesrat. *Verordnung über die Vermeidung und die Entsorgung von Abfällen*; Der Shweizerische Bundesrat: Bern, Switzerland, 2015; p. 46.

96. Suominen, T.; Kunttu, J.; Jasinevičius, G.; Tuomasjukka, D.; Lindner, M. Trade-offs in sustainability impacts of introducing cascade use of wood. *Scand. J. For. Res.* **2017**, *32*, 588–597.

97. Röder, M.; Thornley, P. Bioenergy as climate change mitigation option within a 2 °C target—Uncertainties and temporal challenges of bioenergy systems. *Energy Sustain. Soc.* **2016**, *6*, 1–7.

98. Husgafvel, R.; Linkosalmi, L.; Hughes, M.; Kanerva, J.; Dahl, O. Forest sector circular economy development in Finland: A regional study on sustainability driven competitive advantage and an assessment of the potential for cascading recovered solid wood. *J. Clean. Prod.* **2018**, *181*, 483–497.

99. Näyhä, A. Transition in the Finnish forest-based sector: Company perspectives on the bioeconomy, circular economy and sustainability. *J. Clean. Prod.* **2019**, *209*, 1294–1306.

100. Boisvenue, C.; Running, S.W. Impacts of climate change on natural forest productivity—Evidence since the middle of the 20th century. *Glob. Chang. Biol.* **2006**, *12*, 862–882.

101. Aertsen, W.; Janssen, E., Kint, V.; Bontemps, J.-D.; Van Orshoven, J.; Muys, B. Long-term growth changes of common beech (*Fagus sylvatica* L.) are less pronounced on highly productive sites. *For. Ecol. Manag.* **2014**, *312*, 252–259.

102. Pretzsch, H.; Biber, P.; Schütze, G.; Uhl, E.; Rötzer, T. Forest stand growth dynamics in Central Europe have accelerated since 1870. *Nat. Commun.* **2014**, *5*, 4967. [PubMed]

103. Etzold, S.; Ferretti, M.; Reinds, G.J.; Solberg, S.; Gessler, A.; Waldner, P.; Schaub, M.; Simpson, D.; Benham, S.; Hansen, K.; et al. Nitrogen Deposition is the most Important Environmental Driver Associated to Continental-scale Growth of Pure, Even-aged, Managed European Forests. *For. Ecol. Manag.* **2020**, *458*, 117762.

104. Vittoz, P.; Cherix, D.; Gonseth, Y.; Lubini, V.; Maggini, R.; Zbinden, N.; Zumbach, S. Climate change impacts on biodiversity in Switzerland: A review. *J. Nat. Conserv.* **2013**, *21*, 154–162.

105. Herzog, F.; Seidl, I. Swiss alpine summer farming: Current status and future development under climate change. *Rangel. J.* **2018**, *40*, 501.

106. Rigling, A.; Etzold, S.; Bebi, P.; Brang, P.; Ferretti, M.; Forrester, D.; Gartner, H.; Gessler, A.; Ginzler, C.; Moser, B.; et al. How much Drought can our Forests Endure? Lessons from Extreme Drought Years. In Proceedings of the Forum fur Wissen 2019—Lernen aus Extremereignissen, Davos, Switzerland, 24 April 2019; pp. 39–51.

107. Lindner, M.; Maroschek, M.; Netherer, S.; Kremer, A.; Barbati, A.; Garcia-Gonzalo, J.; Seidl, R.; Delzon, S.; Corona, P.; Kolström, M.; et al. Climate Change Impacts, Adaptive Capacity, and Vulnerability of European Forest Ecosystems. *For. Ecol. Manag.* **2010**, *259*, 698–709.

108. Seidl, R.; Rammer, W.; Lexer, M.J. Climate change vulnerability of sustainable forest management in the Eastern Alps. *Clim. Chang.* **2011**, *106*, 225–254.

109. Die Bundesversamulung der Schweizerischen Eidgenossenschaft. Bundesgesetz über den Wald (Stand am 1. Januar 2017). 1991. Available online: https://scnat.ch/de/uuid/i/a15ef764-3704-53fd-b1df-99391ccd63b5-Der_Wintersturm_Burglind%2FEleanor_in_der_Schweiz (accessed on 18 November 2020).
110. Die Bundesversamulung der Schweizerischen Eidgenossenschaft. *Bundesgesetz über die Raumplanung (Raumplanungsgesetz, RPG)*; Die Bundesversamulung der Schweizerischen Eidgenossenschaft: Bern, Switzerland, 2018; p. 26.

Publisher's Note: MDPI stays neutral with regard to jurisdictional claims in published maps and institutional affiliations.

Review

Improving 'Lipid Productivity' in Microalgae by Bilateral Enhancement of Biomass and Lipid Contents: A Review

Zahra Shokravi [1,*], **Hoofar Shokravi** [2], **Ong Hwai Chyuan** [3], **Woei Jye Lau** [4,5],
Seyed Saeid Rahimian Koloor [6], **Michal Petrů** [6] and **Ahmad Fauzi Ismail** [4,5]

[1] Department of Microbiology, Faculty of Basic Science, Islamic Azad University Tehran-Markazi, Tehran 1477893855, Iran
[2] School of Civil Engineering, Faculty of Engineering, Universiti Teknologi Malaysia, Johor Bahru 81310, Malaysia; shoofar2@graduate.utm.my
[3] School of Information, Systems and Modelling, Faculty of Engineering and Information Technology, University of Technology Sydney, Sydney 2007, Australia; hwaichyuan.ong@uts.edu.au
[4] School of Chemical and Energy Engineering, Faculty of Engineering, Universiti Teknologi Malaysia, Johor Bahru 81310, Malaysia; lwoeijye@utm.my (W.J.L.); afauzi@utm.my (A.F.I.)
[5] Advanced Membrane Technology Research Centre (AMTEC), Universiti Teknologi Malaysia, Johor Bahru 81310, Malaysia
[6] Institute for Nanomaterials, Advanced Technologies, and Innovation (CXI), Technical University of Liberec (TUL), 461 17 Liberec, Czech Republic; s.s.r.koloor@gmail.com (S.S.R.K.); michal.petru@tul.cz (M.P.)
*   Correspondence: shokravi.zahra@gmail.com

Received: 24 September 2020; Accepted: 27 October 2020; Published: 31 October 2020

Abstract: Microalgae have received widespread interest owing to their potential in biofuel production. However, economical microalgal biomass production is conditioned by enhancing the lipid accumulation without decreasing growth rate or by increasing both simultaneously. While extensive investigation has been performed on promoting the economic feasibility of microalgal-based biofuel production that aims to increase the productivity of microalgae species, only a handful of them deal with increasing lipid productivity (based on lipid contents and growth rate) in the feedstock production process. The purpose of this review is to provide an overview of the recent advances and novel approaches in promoting lipid productivity (depends on biomass and lipid contents) in feedstock production from strain selection to after-harvesting stages. The current study comprises two parts. In the first part, bilateral improving biomass/lipid production will be investigated in upstream measures, including strain selection, genetic engineering, and cultivation stages. In the second part, the enhancement of lipid productivity will be discussed in the downstream measure included in the harvesting and after-harvesting stages. An integrated approach involving the strategies for increasing lipid productivity in up- and down-stream measures can be a breakthrough approach that would promote the commercialization of market-driven microalgae-derived biofuel production.

Keywords: biofuel; cultivation strategy; lipid productivity; phycoprospecting; nutrient starvation harvesting; after harvesting; two-stage cultivation

---

1. Introduction

The age of inexpensive fossil fuels is ending. The rapid increase in the world's population and the rising demand for energy are global challenges that have presented themselves over the past couple of centuries [1–3]. However, fossil fuels are not renewable, and their resources are depleting day by day [4–6]. Among the potential energy sources, microalgae-derived biofuels are considered a comparable alternative for fossil fuels [5,7]. Microalgae is a potential feedstock for a number of applications such as the production of animal feed, value-added products, and biofuels [8,9]. However, the possibility of an economically sustainable microalgae feedstock production process is technology-driven, not commercially driven [10–12].

The main economic drawback cited in the literature is that algae species display two conflicting features: high biomass production with low lipid accumulation, or high lipid accumulation with low biomass production [13–15]. Under favorable growth conditions, the oil content of microalgae species typically is between ~10 and 30% dry weights. Meanwhile, there are algae species that produce higher lipid content (56% in *Nannochloris* sp., 80% in *Schizochytrium* sp.). However, the growth rate of such oleaginous species is often slow [16,17]. Alternatively, algae species such as *Scenedesmus* sp. and *Chlorella* sp. with a higher growth rate relatively possess lower lipid content [11,18]. It is commonly known that, for obtaining the best economic scenario, an optimal balance between lipid content and cell growth is required; because culturing either many cells with low lipid content or few cells with high lipid content will not lead to economically sustainable microalgae-derived biofuel production [13,17,19].

To overcome this obstacle, wide-ranging efforts are being made to improve biomass and lipid production in upstream and downstream stages. The upstream measure involves screening appropriate microalgae strain and further improvement of those 'platform algal species' by genetic manipulation to develop new organisms with higher lipid productivity. Furthermore, it is well established that the implementation of an efficient cultivation system that can deal with the contrast between lipid and biomass production is important in enhancing lipid productivity. Establishing the strategies that provide the best performance in the down-stream stage such as harvesting and after-harvesting stages can also provide an additional approach for promoting the lipid productivity of algae strains [20].

Screening for local oleaginous microalgae "bio-prospecting" is the first step in the optimization of the feedstock production process. One of the most important criteria for screening algae strains is lipid productivity (based on lipid contents and growth rate) [21]. It is commonly known that selecting the fast-growing oleaginous algal species would translate directly to an overall feedstock production process [21]. In this regard, researchers have focused their efforts on the screening of microalgae strains with higher biomass and lipid productivities. For instance, in a study, four oleaginous microalgae were investigated for biofuel production. Two of them, *Monoraphidium dybowskii* Y2 and *Chlorella* sp. L1 were found to produce the highest lipid content (32.45, 35.06 mg L^{-1} day^{-1}) and biomass yield (106.61, 137.13 mg L^{-1} day^{-1}) when cultured in photobioreactor (PBR) [22]. In another screening program, Přibyl et al. [23] evaluated the potential of 10 strains of *Parachlorella* and *Chlorella* for lipid and biomass production; among them, the strains *C. vulgaris* CCALA 256, with a biomass density of 5.7 g L^{-1}, and overall lipid content of 30% dry weight, was the most promising strain for biofuel production when cultured in a PBR. Bioprospecting requires high throughput isolating procedures to screen local microalgae strain for biofuel production. However, conventional approaches utilized for strain selection mostly rely on complicated procedures that are labor intensive and time consuming [21,24]. In this regard, Kim et al. [25] studied the novel advances in the development of microfluidic systems for microalgae biotechnology. Similarly, Challagulla et al. [26] reviewed the application of the nuclear magnetic resonance spectroscopy, fluorescent lipid-soluble dyes, Raman spectroscopy, near-infrared spectroscopy, and Fourier transform techniques for analyzing lipid contents in microalgae. It was stated that some technologies such as spectroscopy, flow cytometry and microfluidic can be used alone or in combination with other strategies for the screening of strains with a high growth rate and lipid contents [26–29].

Genetic engineering has previously presented a promising approach for research in different scientific areas [30–33]. As is the case in other research fields, genetic engineering provides an alternative approach to bypass the controversial relationships between lipid accumulation and cell growth [34]. A number of the molecular studies focus on lipid metabolism engineering by over-expression of the enzymes involved in lipid synthesis or suppressing the competitive pathways for enhancing lipid content without comprising the growth rate. For instance, in an investigation, after inactivation of the specific multiuse enzymes acyltransferase/lipase/phospholipase, a mutant strain of *Thalassiosira pseudonana* showed 3.5 times more lipid content without decreasing growth rate [35]. In another study, Nguyen et al. [36] reported that the suppression of the gene involved in fatty acid (FA) degradation of *Chlamydomonas* mutant, Cre01.g 000300, could increase lipid content without impacting the growth rate. Up to date, a number of review articles have been published on genetic engineering techniques for promoting the lipid productivity of algae species. For instance, Park at al. [37] surveyed the strain improving strategies such as genetic engineering, random mutagenesis, and metabolic engineering pathways to promote lipid productivity of microalgae; and stated the combination of the appropriate tools and right targets can improve algal species that would promote the commercialization of market-microalgae-derived biofuel production. In similar investigations, Chen et al. [14] proposed that the lipid production efficiency in microalgae can be increased by integrating stress tolerance manipulation strategies with genetic engineering approaches.

It is commonly known that microalgae cultivation is the most important stage in microalgae biofuel production because the quantity and quality of the produced feedstock will strongly depend on this stage [19]. An ideal cultivation strategy would enable algae strains to grow rapidly with a synchronized increase in lipid content. However, cultivation conditions for cell growth typically differ from that required for lipid production. Generally, the microalgae species cultured under favorable conditions produce large amounts of biomass but with lower lipid yield [9]. Thus, for increasing lipid content in algae, additional approaches, such as applying different stresses during the biomass production process have been proposed. But such stressful conditions often have a negative impact on the microalgae growth rate, leading to a decrease in the desired product yield. Thus, obtaining the best economic scenario will require an optimal balance of lipid content and cell growth [13,19]. In this sense, the two-phase system has been proposed as a win-win strategy to overcome the trade-off effect between biomass and lipid yield. In the two-step cultivation system, a nutrient-rich growth medium is used in the first step to obtaining maximal biomass production. After an adequate concentration of algal biomass is produced, the medium condition changes into a stress induction condition in the second step. In an investigation, Xia et al. [38] developed a salinity-based two-phase cultivation for *Scenedesmus obtusus* XJ-15. In the first phase, the biomass productivity was increased from 139.4 to 212.1 mg L^{-1} day^{-1}. In the second phase, lipid content was increased from 26.1% to 47.7%. In another investigation, Yun et al. [39] proposed a two-phase system for *N. oleoabundans* HK-129. The process resulted in a 40 and 60% increase in algal biomass and lipid content, respectively. Additionally, a number of review articles have been done as part of an effort to study the efficiency of various cultivation strategies in microalgae. One such investigation by Ho et al. [19] studied the efficiency of various cultivation systems for increasing lipid productivity in microalgae and stated two-stage and semi-continuous strategies can increase lipid content without impacting the growth rate. Nagappan et al. [40] compared lipid and carbohydrate productivities of two-stage strategies with single-stage systems. Aziz et al. [41] highlighted the potential of a two-stage culture strategy for simultaneous lipid and biomass production and modified the pre-harvesting stage to promote the economic viability of this strategy. In one of such studies, Bhatia et al. [42] discussed different types of wastewater and summarized the recent approaches in algal cultivation and harvesting technologies from wastewater. Table 1 shows a list including studies about increases in lipid and/or growth rate in microalgae, brief findings of which are given presently.

Table 1. Summary of literature about the enhancing lipid and/or biomass production in microalgae.

Year	References	Genus and Species	Upstream Approaches			Downstream Approaches	
			Strain Selection	Genetic Approach	Cultivation	Harvesting	After Harvesting
2012	Přibyl et al. [23]	*C. vulgaris*	✓	✗	✗	✗	✗
2013	Borowitzk et al. [43]	*	✓	✗	✓	✗	✗
2013	Xia et al. [38]	*S. obtusus*	✗		✓	✗	✗
2013	Trentacoste et al. [35]	*T. pseudonana*	✗	✓	✗	✗	✗
2014	Ho et al. [19]	*	✗	✓	✓	✗	✗
2015	Singh et al. [44]	*	✗	✓	✗	✗	✗
2016	Challagulla et al. [26]	*	✓	✗	✗	✗	✗
2017	Chen et al. [14]	*	✗	✓	✓	✗	✗
2017	Chu [45]	*	✓	✓	✗	✗	✗
2017	Kim et al. [25]	*	✓	✗	✗	✗	✗
2017	Chung et al. [46]	*	✗	✓	✗	✗	✗
2018	Sharma et al. [34]	*	✗	✓	✗	✗	✗
2018	Yun et al. [39]	*N. oleoabundans*	✗	✗	✓	✗	✗
2018	Shin et al. [47]	*	✓	✗	✓	✗	✗
2018	Sajadi et al. [16]	*	✓	✗	✗	✗	✗
2019	Nagappan et al. [40]	*	✗	✗	✓	✗	✗
2019	Piligaev et al. [48]	*	✗	✗	✓	✗	✗
2019	Park et al. [37]	*	✗	✓	✗	✗	✗
2019	Menegazzo et al. [49]	*	✗	✗	✗	✓	✗
2020	Nguyen et al. [36]	*Chlamydomonas* sp.	✗	✓	✗	✗	✗
2020	Poh et al. [50]	*Chlorella* sp.	✗	✗	✗	✗	✓
2020	Aziz et al. [41]	*	✗	✗	✓	✗	✗
2020	Bhatia et al. [42]	*	✗	✗	✓	✓	✗

* The reference is a review article and a large variant of species and strains are studied. A tick mark (✓) and cross mark (✗) indicates that the variable is and is not included in the study, respectively.

Among the considered approaches, establishing the strategies that provide the best performance in the harvesting stage without having qualitative damages to microalgae biomass and/or lipid content is also important in the successful implementation of the feedstock production process [51]. For instance, Liu et al. [52] applied magnetite nanoparticles (nano-Fe_3O_4 coated) for the harvesting of *C. pyrenoidosa* and *S. obliquus* strains, and reported that this harvesting method did not reduce the lipid content in these species. Very few studies have been performed on harvesting and after-harvesting stages, although these stages should be taken into consideration for improving lipid productivity in the feedstock production process [50,51]. In one of such studies, Menegazzo et al. [49] reviewed the culture conditions for various algae species and their influence on the separation of microalgae biomass, lipid content, including biomass thickening methods, and methods of biomass depletion, the methods of cellular disruption and lipid extraction.

The above literature review shows that most of the studies provide only a partial picture of the bilateral increase in lipid productivity throughout the cultivation or genetic engineering stages. Meanwhile, there are review articles that focused on more aspects of the biomass production process. Studies such as that by Chu [45] present an overview of strain selection, genetic engineering, and cultivation without depth information about different cultivation strategies. Chen et al. [14] focused on recent advancements in the lipid enhancement process in the cultivation and genetic engineering stages. However, these articles are not containing all relevant information about increasing lipid productivity throughout the feedstock production process, and only scattered information is accessible (Table 1). To the best of the authors' knowledge, no current review paper contains the findings of promoting lipid productivity in strain selection, genetic manipulation, cultivation, harvesting, and after harvesting stages. The purpose of this review is to provide an overview of the recent advances and novel approaches in promoting lipid productivity in the biofuel production process from strain selection to after-harvesting stages.

The first section of the article offers a discussion about the importance of bilateral biomass and lipid production in biofuel production and its potential in increasing microalgae economic feasibility of microalgae-derived biofuel production. In the following section, a bilateral increase in biomass/lipid production is investigated in upstream measures, including strain selection and genetic engineering stages. Finally, promoting lipid productivity will be discussed in the downstream measure included in the harvesting, and after-harvesting stages (Figure 1).

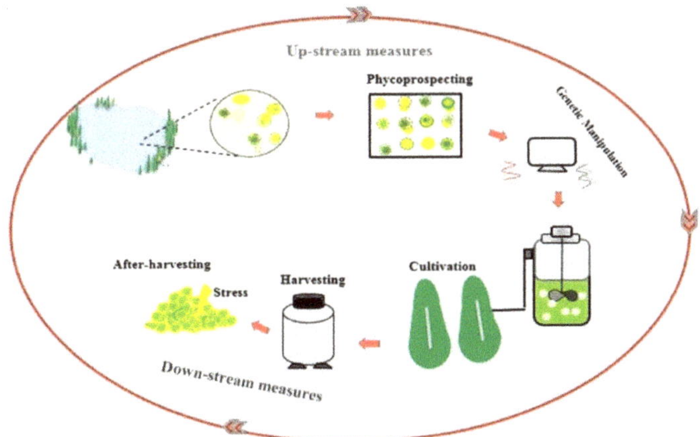

Figure 1. Enhancing lipid productivity in the feedstock production process.

2. Bilateral Improvement in Biomass and Lipid Productivities

Microalgae are potential biofuel feedstock, owing to its rapid growth rate and ability to produce value-added lipids [16]. The lipid content of microalgae can be increased when the microalgae cells are subjected to stressful conditions such as culturing under a nutrient limitation or environmental stress [17]. However, such stresses often negatively affect algal growth, leading to decreased lipid productivity [47,53,54]. In an investigation, the effect of stressful conditions on biomass and lipid production of *Chaetoceros muelleri* was studied [55]. Under nitrogen-deficient conditions, the lipid content was elevated from 23% to 46%. However, on the other side, biomass productivity was decreased from 0.19 to 0.12 g L^{-1} day^{-1} [44]. Similarly, in another study, the impact of nutrient limitation on lipid and biomass yield of *Scenedesmus destricola* and *Chorococcum* spp. was investigated [56]. Under nitrogen-limited conditions, the lipid yield was increased from 48% to 54% for *Scenedesmus destricola* and 31.6% to 40.7% for *Chlorococcum nivale*. The biomass productivity of *Scenedesmus destricola* and *Chlorococcum nivale*, however, was decreased from 0.48 to 0.38 g L^{-1} day^{-1} and from 0.40 to 0.38 g L^{-1} day^{-1}, respectively [44].

It is well established that increasing lipid content, without impacting the growth rate or increasing both, holds the key to obtaining economic viability of algae-based biofuel production [38]. The mathematical analyses of Yu et al. [57] showed how both biomass productivity and lipid contents are vital in determining the biofuel production cost. In their first analysis of *Chlorella vulgaris* CCAP 211/11B and *Chlorella vulgaris* F&M-M49 of similar lipid content and cell size, *Chlorella vulgaris* F&M-M49 incurred lower production costs because of its higher biomass yield. In other analyses of *Nannochloropsis* sp. and *Nannochloropsis* sp. F&M-M27 with similar biomass productivity and cell size, *Nannochloropsis* sp. F&M-M27 reduced the cost by 10–20%, owing to its 25% higher lipid content [47].

Conventional stress induction approaches seem to be useful for increasing algal lipid content but often fail to increase lipid productivity as the enhancing lipid content happens at the cost of biomass. Lipid productivity as expressed in Equation (1) usually reported in the unit of g lipid m^{-2} day^{-1} or g lipid L^{-1} day^{-1}.

$$Lipid\ Productivity\ =\ \mu Q \tag{1}$$

where Q is the amount of the microalgal yield per unit area of culture (m^2) or volume (L) and μ is the specific growth rate (day^{-1}) [37,43]. Therefore, microalgae-based biofuel production success is strongly influenced by two mutually exclusive factors, i.e., lipid content and growth rate. High lipid content is essential to decrease the processing costs per unit of biomass products, and a high growth rate is required to increase yield per unit culture area. Generally, lipid accumulation in algae typically is not concurrent with a fast growth rate, which is the potential conflict in the production process [58]. To achieve the economically feasible cost at a necessary production scale, it is reasonable to develop an optimal balance among upstream and downstream technologies to reduce overall process costs (Figure 2).

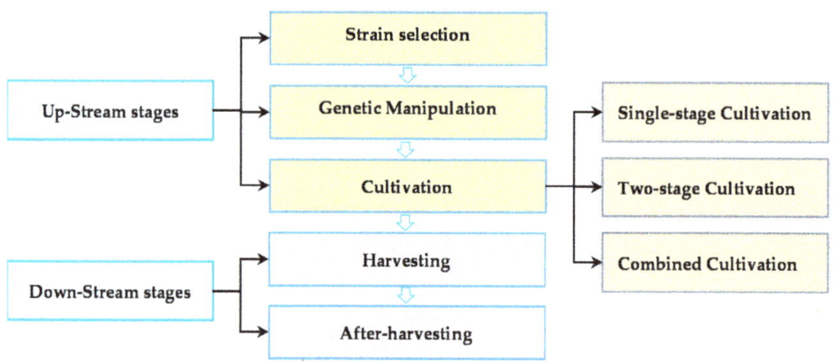

Figure 2. Diagram of the microalgae feedstock production procedure.

Two trends have been proposed in biofuel-related algae studies for achieving high lipid productivity. The first trend focuses on enhancing lipid content without decreasing biomass yield, which is often achieved by genetic manipulation approaches. For instance, the co-expression of five acyltransferases from yeast in *Chlorella minutissima*, elevated lipid yield by two times without decreasing the growth rate [59]. The other trend has the aim of increasing biomass and lipid yields simultaneously that generally can be attained by applying a suitable cultivation strategy. For instance, Moussa et al. [60] designed a hybrid cultivation system for *Picochlorum* sp., in which maximal biomass productivity (0.427 g L^{-1} day^{-1}) was obtained by the continuous culture at a dilution rate of 0.6 day^{-1}, and high lipid contents, ranging from 499 to 698 g kg^{-1} dry weight (DW), were obtained under different nitrogen sources in the second step. In another study, San Pedro et al. [61] developed a two-stage cultivation method for *Scenedesmus* sp. In this cultivation method, microalgal cells were maximal biomass productivity of 0.49 g L^{-1} day^{-1} was attained with a nitrate concentration of 8.0 mM at a dilution rate of 0.42 1/day. A high increase in lipids content, 73.1% of overall lipids, was obtained in the second phase under nitrogen-depletion condition.

3. Upstream Measure

Oleaginous microalgae are potential cell factories for viable biodiesel production. It possesses the inherent ability to accumulate value-added products such as lipids and exhibits a high growth rate [20]. Commercially sustainable microalgae-based biofuel production requires screening local oleaginous species, which can be further promoted by strain engineering strategies [37,62]. Therefore, the upstream technologies mainly include three aspects. Firstly, screening appropriate algal species that are characterized by rich in lipid and fast-growing rate. Secondly, advanced molecular approaches can be manipulated so that produce the strains with high biomass and lipid productivity. Finally, selecting and establishing an efficient cultivation system that can deal with the contrast between lipid and biomass production.

3.1. Phycoprospecting

As estimated, there are 1 to 10,000,000 alga species on Earth; however, most of the on-going studies have been focused on a limited number of microalgae strains. Microalgae are characterized based on high lipid content and fast growth rate. But all of the algae species are not regarded as the best lipid producers [20,25,63]. Although documented in different studies about increasing the microalgal lipid productivity via various approaches, the potential limitation cannot be overcome if the screened algal species are not apt for biodiesel production [17,47]. Identifying best performing microalgae strains through 'phycoprospecting' can be a promising approach to obtain strains with higher lipid productivity for achieving commercially sustainable biofuel production [47,64].

In this respect, the fundamental requirement of microalgae-based biodiesel production is selecting appropriate strains with a high level of lipid accumulation and high biomass productivity [65]. Additionally, the optimal microalgae strains for biofuel production should have a number of important features: an optimal composition of lipid profiles; high cell density during cultivation; high carbon dioxide absorption rate; lack of need for expensive nutrients during cultivation process; resistance to temperature fluctuations; co-production of valuable by-products and short production cycle [17].

Besides, local algal species are more preferred for feedstock production purposes as they have better adaptability to environmental situations prevailing in a specific geographical location [15]. Several programs have focused on the isolation of the appropriate indigenous algal strains to enhance lipid and biomass productivities. Screening of indigenous microalgae strains for desirable characteristics or 'phycoprospecting' is important in determining potential strains for biodiesel production [65,66]. For example, an isolation project under the National Alliance for Advanced Biofuels and Bioproducts (NAABB) has successfully screened novel potential 'platform strains' with fast growth rates and high lipid content [66].

Bio-prospecting requires high throughput and rapid isolating procedures to screen new strains that are adapted to the production location [65]. Several techniques have been described, including single-cell isolation using serial dilutions, micromanipulation, atomized cell spray, and gravimetric separation [67,68]. However, such traditional microalgae isolation methods are time consuming and require further screening of axenic cultures to determine lipid and growth productivities [27,69,70]. Technologies such as automated processes including robotics, flowcytometry, and new strategies such as microfluidics and deep sequencing are being developed to facilitate the isolation and characterization of microalgae strains [70]. A combination of modern and conventional techniques may screen the suitable algal strain for biofuel production. In a study, flowcytometry technique coupled to cell sorting strategies has led to a rapid selection of strains with high lipid contents and fast growth rates [69,70]. In a study, Huang et al. [65] innovated a novel direct sampling technique that an enrichment strategy was coupled with a capillary aided sampling procedure. This approach sped up the isolation of desirable strains for both rapid growth and high lipid productivities. Kim et al. [71] innovated a droplet microfluidics analysis platform that is possible to study the differences in lipid and biomass productivities of *Chlamydomonas reinhardtii* under different nitrogen conditions. Table 2 highlights the techniques that have been applied for screening microalgae with lipid content/growth rate.

Table 2. Some techniques have been applied for screening microalgae with lipid content/growth rate.

Genus and Species	Approach	Comments	Ref.
Several microalgae strains	Inverted fluorescence microscope	Analyzes growth and lipid content	[65]
Chlorococcum littorale	Fluorescence assisted cell sorting	detect microalgae cells with high lipid content	[69]
Phaeodactylum tricornutum	microfluidic cytometer	measures lipid accumulation and photosynthetic efficiency	[28]
Dunaliella salina	Fourier transform infrared spectroscopy	Examines the growth and lipid yield	[29]
Chlamydomonas reinhardtii mutants	Droplet microfluidics-based screening	Analyze growth and lipid content in populations	[25]
Chlorococcum littorale	Fluorescence assisted microalgae screening	Screen algal cells with high lipid content under nitrogen deficiency	[69]

3.2. Molecular Approaches

Microalgae combine biotechnological properties of microbial cells (the ability to accumulate metabolites and fast growth) with typical characteristics pertaining to higher plants (simplicity of nutritional requirements and efficient oxygenic photosynthesis). This specific combination establishes the basis of biotechnological approaches for increasing lipid productivity in microalgae [72]. Oleaginous microalgae have attracted significant interest in the production of biofuel owing to its capacity in producing large amounts of lipid and fast growth rate [73]. However, lipid accumulation in microalgae is not conducive to a high growth rate, which is the fundamental limitation in the feedstock production process. To overcome such a drawback, genetic engineering provides an alternative approach to bypass the controversial relationships between lipid accumulation and growth rate [74]. It was reported that genetic engineering strategies that increase lipid accumulation without compromising growth rate could reduce production cost and fortify the economic sustainability of algae-based biofuels production [75].

So far, most of the molecular investigations focus on lipid metabolism engineering by over-expression of the enzymes involved in lipid synthesis or suppressing the competitive pathways in lipid or biomass production [14]. Meanwhile, genetic engineering of the genes involved in stress tolerance mechanisms exhibits a significant potential in improving lipid productivity [76]. The integration of these strategies may provide a potential approach for sustainable microalgae-based biofuel production at a competitive cost.

3.2.1. Lipid Biosynthesis Pathway

Triacylglycerol and FA synthesis included a series of biochemical reactions medicated by different enzymes. These enzymes' over-expression would lead to enhancing the enzyme activity and thereby effectively triggers lipid accumulation [77]. Acetyl-CoA carboxylase (ACCase) is one of the most exploited enzymes for enhancing lipid accumulation in microalgae. As described by previous reports, the overexpression of ACCase had less effect on lipid production [45]. However, the synchronized overexpression of malic enzyme (ME) with a subunit of ACCase (*accD*) was successful in increasing the lipid productivity of *Dunaliella salina* [78]. Similarly, the overexpression of malic enzyme is reported to enhance the lipid production of *Phaeodactylum tricornutum* by 2.5 folds without a negative impact on the growth rate [79].

One of the advances in this area was obtained by the co-expression of five acyltransferases from yeast in *Chlorella minutissima*, which elevated lipid production by 2 times without compromising the growth rate [59]. In addition, the overexpression of several other enzymes, such as acetyl-CoA synthase, phosphoenolpyruvate carboxylase, pyruvate dehydrogenase, glycerol kinase, and NAD (H) kinase has been reported to increase lipid accumulation without compromising the growth rate. A summary of the molecular studies employed for increasing lipid content without sacrificing biomass production is listed in Table 3.

Table 3. Overviews of molecular approaches employed for overexpression/suppression lipid content without comprising growth rate.

Genus and Species	Approach	Gene	Phenotypic Changes	Ref.
Phaeodactylum tricornutum	Suppression	Pyruvate dehydrogenase kinase	Improves lipid (up to 82%)/slight decrease in growth	[80]
Fistulifera solaris	Overexpression	Glycerol kinase	Improves lipid/biomass productivities	[81]
Chlorella pyrenoidosa	Overexpression	NAD (H) kinase	Enhance in lipid accumulation by 110.4%/without decreasing growth rate	[82]
Chlamydomonas reinhardtii	Suppression	Phosphoenol pyruvate carboxylase	Improves lipid content (14–28%)	[83]
Phaeodactylum tricornutum	Overexpression	Malic enzyme	Improves lipid content (2.5-fold) but did not affect the growth rate	[79]
N. oleoabundans	Overexpression	lysophosphatidyl-acyltransferases / Glycerol-3-phosphate acyltransferase / diacylglycerol acyltransferase	Increase in lipid content (52% and 42%, respectively) without decreasing growth rate	[84]
Chlorella minutissima	Overexpression	Co-expression of five acyltransferases	Improves lipid content (up to 2-fold) but did not affect the growth rate	[59]

Among the considered strategies, suppressing competitive pathways such as lipid and carbohydrate catabolism is another approach for improving lipid productivity [85]. Carbohydrate metabolism is the most important pathway in microalgae for carbon storage and starch production. Knocking down the starch metabolism pathway may result in the carbon flow towards lipid synthesis [86]. However, the inhibition of the genes involved in starch biosynthesis may result in a reduced growth rate, consequently decreasing lipid productivity [87,88].

The suppression of lipid catabolism, particularly the enzymes that catalyze the FA release, is another promising approach for increasing algal lipid productivity [35,88,89]. Lipid catabolism promotes membrane reconstruction by providing acyl groups, which is necessary for membrane reorganization of the photosynthetic system in microalgae [36,90]. However, some studies have indicated that, unlike the suppression of carbohydrate catabolism pathways, the inhibition of genes involved in lipid catabolism may have less influence on growth rate [35,89]. Therefore, enzymes involved in algal lipid catabolism have been considered as a potential alternative for a simultaneous increase in biomass and lipid yield. In an investigation, a mutant strain of *Thalassiosira pseudonana* showed 3.5 times more lipid synthesis after inactivation of the specific multiuse enzymes acyltransferase/lipase/phospholipase [35]. In another study, Nguyen et al. [36] reported that the suppression of the gene involved in FA degradation of *Chlamydomonas* mutant, Cre01.g 000300, could increase lipid content without impacting the growth rate.

3.2.2. Molecular Approaches for Modulation the Stress-Related Mechanisms

Lipid biosynthesis pathway in algae is a multi-step reaction, catalyzed by an enzyme complex [63]. Under optimal conditions, algae cell growth is endorsed by increased transcription and translation processes, which led to high biomass productivity [44]. However, under nutrient starvation conditions the microalgae growth is constrained, as most of the anabolic machinery is retarded. Therefore, identification of the behavior and mechanism of these enzymes under various environmental conditions is important in improving the lipid productivity of microalgae strains [14]. Wan et al. [88] studied the impacts of iron concentrations on the lipid yield of *Chlorella sorokiniana*. The expression of *acc1*, *accD* and *rbcL* genes are up-regulated at higher concentrations of iron, leading to greater lipid yield, without negatively affecting the growth rate. Similarly, Fan et al. [82] showed that the overexpression of the AtNADK3 of *Chlorella pyrenoidosa* considerably increased the lipid content with no adverse impact on the growth rate.

Additionally, manipulation of stress-responsive promoters is considered a potential approach to increase lipid production without negative impacts on growth rate [14]. For example, overexpression of the diacylglycerol acyltransferases gene controlled by the phosphorus limitation-inducible promoter has promoted lipid production by 2.5 folds in an engineered strain of *Chlamydomonas reinhardtii*, when compared to the control group [91]. Further heterologous expression of this construct in *Nannochloropsis* sp. NIES-2145 under phosphorus limitation also resulted in higher lipid yield (1.7 times more than the wild strain) [92].

Key transcription factors (TFs), as well as the enzymes and promoters, can be targeted for genetic engineering to achieve high lipid productivity [14]. Overexpression of the TFs involved in the regulation of lipid biosynthesis pathways can divert the metabolic flux toward lipids accumulation. Therefore, identification of TF-encoding genes and their subsequent manipulation in their hosts would be an efficient genetic approach for developing the robust microalgae strains [14]. Successful reports have emerged to verify the important roles of different TFs in enhancing the lipid yield without decreasing the growth rate (Table 4). For instance, TF GmDof4 expression from *Glycine max* in *C. ellipsoidea* contributed to enhancing the lipid accumulation from 46% to 53% with no negative impacts on growth rate [93]. In other investigations, the manipulation of CHT7 [94] and PSR1 [95] elevated lipid production in *C. reinhardtii* without comprising the growth rate. Similarly, Ajjawi et al. [96] studied the transcriptional profiling of *N. gaditana* under nutrient limitation conditions. They identified 20 efficient TFs for lipid accumulation in *N. gaditana*, and performed insertional knockout on 18 of these 20 TFs by CRISPR/Cas9 reverse-genetics pipeline.

Table 4. Some genetic studies are employed for transcription factors (TFs) in different microalgae species.

Species	Transcription Factors	Purpose of Modification	Ref.
C. reinhardtii	Compromised Hydrolysis of Triacylglycerols (CHT7)	Manipulation of CHT7 TF increased lipid productivity	[94]
C. reinhardtii	PSR1	Manipulation of PSR_1 TF increased growth rate (two-fold or more) and an increase in lipid content	[95]
N. salina	NsbHLH2	Overexpression of NsbHLH2 increased biomass productivity (509.3 mg L^{-1})/lipid content (9.96% DW)	[97]
N. salina	Basic leucine zipper (bZIP)	Under the N-deprivation conditions, transformants showed an increase of up to 88% and 39% in lipid content and biomass productivity, respectively.	[98]
C. ellipsoidea	GmDof4	Overexpression of GmDof4 increased the lipid content by 52.9% but did not affect the growth rate	[93]

Table 4. *Cont.*

Species	Transcription Factors	Purpose of Modification	Ref.
D. bardawil	WRKY	WRKY up-regulates carotenogenic genes to increase carotenoid and is important in adaptation to abiotic stress	[99]
P. tricornutum	RING-GAF-Gln	RGQ$_1$ is involved in early nitrogen starvation	[100]
N. gaditana	Zn(ii)$_2$Cys$_6$	Doubled the strain's lipid content without decreasing growth rate	[96]
H. pluvialis	Myb	TFs can affect other TFs to enhance astaxanthin/carotene biosynthesis	[101]

3.3. Cultivation Stage

Although biomass/lipid production in microalgae is species-specific, the type of selected cultivation system is important in determining whether a feedstock production system will be economically feasible for biofuel production [44]. An ideal cultivation strategy would enable algae strains to grow rapidly with a synchronized increase in lipid content [19]. Cultivation strategies are categorized to single-stage strategies (e.g., semi-continuous, fed-batch, and continuous), integrated strategies and two-stage cultivation systems (Figure 3). [17,19,102]. These cultivation strategies, with emphasis on enhancing simultaneous biomass and lipid production, are discussed in the next sections.

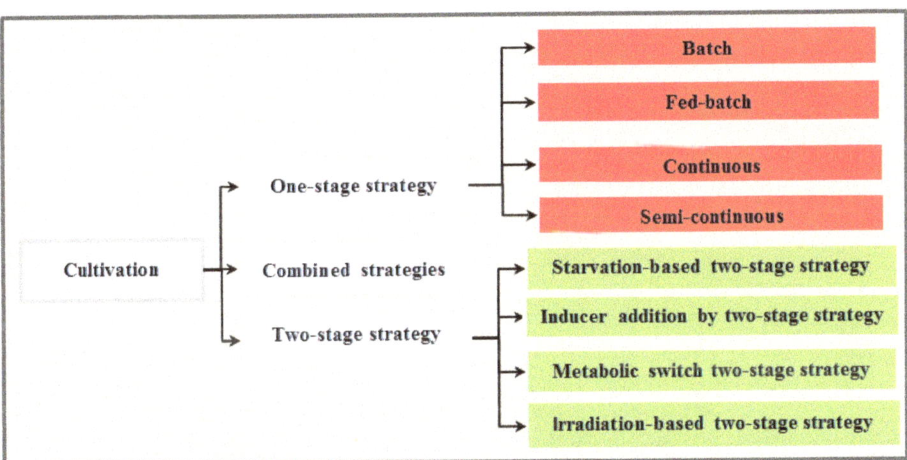

Figure 3. Diagram of the different cultivation systems.

3.3.1. One-Stage Cultivation Strategy

Single-stage cultivation systems are categorized into five main groups: batch, continuous, fed-batch, and semi-continuous. As mentioned above, cultivation systems strongly influence the optimization of algal biomass and lipid production [102]. A comparison of the characteristics of various single-stage strategies indicates that the high lipid productivity can be achieved readily using semi-continuous strategy due to the strong stressful conditions associated with this cultivation system [19].

In contrast to the semi-continuous strategy, fed-batch and continuous cultivation systems are applied when the required nutrient level is available from the beginning of the cultivation process; in such cases, the medium gradually reaches a point of nutrient deprivation, which lead to lower lipid yield [79]. Generally, continuous systems tend to provide negligible lipid content, leading to their dismissal as sustainable platforms for microalgae-based biodiesel production [103]. However, some investigations have shown the possibility of using these cultivation strategies to improve microalgae biomass and lipid productivities through precision nutrient limitation. For instance, simultaneous lipid and biomass production was obtained in continuous chemostat culture when continuous feeding of BG11 media was supplemented with lipid inducers such as sodium chloride and sodium acetate [103]. Accordingly, Wen et al. [104] achieved high biomass and lipid productivity in continuous chemostat mode when the specific nitrate input rate was in the range of 0.78–4.56 mmol g^{-1} day^{-1}. In other investigations, Del Rio et al. [105] cultivated *Pseudokirchneriella subcapitata* in continuous chemostat culture. The highest FA productivity was reported in nitrate concentration ranging from 3 to 5 mM. Some examples of such studies are outlined in Table 5.

Table 5. Lipid/biomass productivities in different algae strain cultivated under a single-stage cultivation strategy.

Microalgae Species	Strategy Adopted	Productivity (mg L^{-1} day^{-1})	Dilution Rate	Ref.
Choricystis minor	Continuous	82	0.014 h^{-1}	[106]
Chlorella minutissima *Dunaliella tertiolecta*	Continuous	1.37 0.91	0.33 day^{-1} 0.42 day^{-1}	[107]
Chlorella pyrenoidosa	Fed-batch Batch	1.45 96.28	388.0 µg L^{-1} h^{-1}	[102]
Neochloris oleoabundans	Continuous	-	-	[108]
Chlorella vulgaris *Chlorella sorokiniana* *Neochloris oleoabundans* *Chlorococcum oleofaciens* *Scenedesmus naegleii* *Scenedesmus dimorphus*	Batch	0.94 0.85 1.31 0.86 0.83 1.11	-	[13]

3.3.2. Two-Stage Strategy

In microalgae, the cultivation conditions for lipid accumulation are different from those needed for biomass production. Thus, how to optimize cultivation conditions to achieve high lipid and biomass and production, is a big challenge in feedstock production [41,109]. In this sense, the two-phase system has been proposed as a win-win strategy to overcome the trade-off effect between biomass and lipid yield [20]. In the two-step cultivation system, a nutrient-rich growth medium is used in the first step to obtaining maximal biomass production. After an adequate concentration of algal biomass is produced, the medium condition changes into a stress induction condition in the second step [19,110]. It was reported, in the two-step system, that the average increase in biomass production is higher than in non-hybrid strategies, 12.5% more than photobioreactors, and 46–74% more than open ponds [41].

Based on various stimuli, two-stage cultivation strategies are classified into five groups: inducer addition by two-stage strategy, starvation-based two-stage strategy, metabolic switch by two-stage strategy, and irradiation-based two-stage strategy [40,41]. The next section of this article discusses various kinds of two-stage cultivation systems based on different stresses are described.

Nutrient Starvation

Nutrient starvation is demonstrated as an efficient approach to enhance lipid accumulation [111,112]. Of many various nutrient starvation strategies, nitrogen deprivation is one of the reliable methods to enhance lipid production in microalgae [16,113]. However, under nutrient depletion conditions, the growth rate is considerably decreased, resulting in lower lipid productivity. Accordingly, the two-stage strategy is proposed to address this issue, in which lipid and biomass production are split into separate steps [114,115]. In nutrient limitation strategy, the duration ratio between nitrogen-replete and deplete phases as well as the initial cell density in the second stage are crucial in optimization algal lipid and biomass production [11,116].

In terms of diatoms, silica deprivation is a preferred stress strategy for enhancing lipid accumulation [109]. The silica-limitation effect is more severe and rapid than nitrogen deficiency in this taxonomy. It was reported that silica deprivation can provide a controllable approach to enhance lipid biosynthesis in the two-step strategy. As an example, silica deprivation enhances the lipid content in diatoms of *Cyclotella cryptica*, *Navicula saprophila* and *Chaetoceros muelleri* up to almost 104%, 110%, and 89%, respectively [16,117].

Due to the efficiency of both nutrient limitation methods, a novel two-step model with nitrogen (N)-silicon(Si)-starvation has been suggested. In this strategy, the co-limitation of silicon and nitrogen not only improved lipid accumulation but also increased biodiesel quality of *Skeletonema costatum* [118].

Other than nitrogen and silicon, phosphorus is also vital for microalgal cell growth. In a number of cases, phosphate deprivation was reported to be more efficient than nitrogen starvation, such as the case in the two-step strategy of *Ankistrodesmus falcatus* where lipid productivity was higher than that in the nitrogen limitation method [119–121]. Additionally, reports on the effects of phosphate starvation on algal lipid productivity, based on the experimental setup, seem to be strain dependent. For instance, the lipid accumulation of *Phaeodactylum tricornotum*, *Pavlova lutheri*, *Chaetoceros*, and *Dunaliella salina* enhanced under phosphorus starvation, but, conversely, under such a condition, lipid content in *Tetraselmis*, *Chlorella*, and *Nannochloris atomis* was decreased. It is indicated that not all algae strains enhance their lipid accumulation during phosphate starvation conditions [55,120,122].

Inducer Addition

Although different inducers such as high salinity, low salinity, halo-alkalinity, and phytohormones could have a positive influence on algal lipid production [123,124], the changes in the culture medium are often associated with decreased cell growth in exchange for lipid production [125]. Therefore, integrating different inducers with a two-stage strategy can be an efficient strategy in optimization algal biomass and lipid productivities. For instance, applying salinity stress in the second phase of the two-step strategy increased the lipid content of *I. galbana* from 24% to 47% [126]. Similarly, inducing salinity stress in the two-step system was found to increase the lipid yield of *Monoraphidium dybowskii* LB50 [127]. It was reported that the salinity-based two-step cultivation strategy not only increases lipid productivity but also promotes the biodiesel quality obtained from the microalgae [40]. For instance, the biodiesel properties of saponification value, cetane number, long-chain saturation factor, and iodine value are considerably promoted in the salinity-based two-step strategy of *Scenedesmus obtusus* XJ-15 compared to one-step strategies [40,128].

A combination of high alkaline salt and pH (halo-alkalinity) can be readily adopted in this cultivation strategy. It was reported that the addition of hydrochloric acid and sodium bicarbonate can control pH in the culture medium in certain types of microalgae [40]. Wensel et al. [129] designed a halo-alkalinity-based two-step strategy in *Chlorella pyrenoidosa* and achieved a high lipid and biomass yield as well as high autoflocculation harvesting efficiency of 64.1%.

As mentioned above, microalgae lipid productivity can be improved by the induction of stressful conditions. However, such stresses often negatively affect the photosynthetic activity and, consequently, decrease the production of desired products [16]. The major reason for the decreased photosynthesis activity is the production of reactive oxygen species (ROS), which impairs the photosynthetic systems [130]. To solve this problem, a number of approaches have been suggested to reduce the oxidative stress caused by stressful conditions and hence increase lipid and biomass production [131,132].

Recent studies on the bicarbonate application in the cultivation process show that not only is it considered as a carbon source, but also act as an oxidative stress mitigator [86]. Under nutrient depletion conditions, the addition of sodium bicarbonate considerably decreased the oxidative stress caused by ROS and promoted the activities of antioxidant enzymes of *Dunaliella salina*, resulting in increased lipid and biomass production [130]. In an investigation, manipulating the iron, nitrate, and carbonate was reported to increase lipid and biomass production of three microalgae *Chlorella* sp., *Scenedesmus* sp., and *Chlamydomonas* sp. [133]

Apart from bicarbonate addition, phytohormones also play an important role in metabolism regulation and growth of microalgae [86,131]. Phytohormone could increase the antioxidant system in the microalgae, which keeps a redox balance state under stressful conditions. The combination of plant hormones with a two-stage culture system may be efficient in increasing the higher biomass and lipid production in microalgae [131,134]. For instance, the integration of a two-stage strategy (heterotrophic-photoautotrophic) with fulvic acid results in an increase of 54–65% in the lipid content of *Monoraphidium* sp. [135]. Table 6 shows some examples of inducer-based two-stage cultivation.

Table 6. Lipid/biomass concentration of growth mediums in different strains of microalgae supplemented with various types of inducers.

Genus and Species	Lipid Productivity $(g\ L^{-1}\ day^{-1})$	Type of Inducer	Ref.
Tetraselmis sp.	0.285	Salinity + Nitrogen starvation	[125]
Chlorella sorokiniana DPK	0.690	Diethyl aminoethyl hexonate(DA-6), Indole-3-acetic acid (IAA) and N starvation	[136]
Chlorella sorokiniana	0.502 0.494	IAA cytokinin- kinetin (K)	[132]
Scenedesmus obtusus XJ-15	0.607	Salinity stress	[38]
Nannochloropsis oculata	0.324	Salinity stress	[137]
Dunaliella salina KSA-HS022	0.565	Salinity stress	[138]
Chlamydomonas reinhardtii	0.109	Combination of NaCl/CaCl$_2$	[139]
Chlorella vulgaris	0.800	Salinity + Nitrogen starvation	[140]

Metabolic Switch

Microalgae can adopt different trophic modes, including photoheterotrophic, heterotrophic, and mixotrophic, based on energy and the available carbon source [40]. Algal biomass is generally produced through a photoautotrophic culture where microalgae can convert water and carbon dioxide into feedstock through the photosynthesis process [141]. Among all the required cultivation conditions, adequate light is recognized as a crucial factor for photoautotrophic conditions. Generally, insufficient light penetration, resulting from mutual shading, is the main limiting factor for microalgae cultivation under an autotrophic mode [142]. Thus, transferring the microalgae growth culture into the second phase under mixotrophic or heterotrophic conditions can reduce the culture's need for sunlight [17,143].

Microalgae cells are typically grown in a photoautotrophic mode in the first phase, and then transfer the cultured algal biomass into a heterotrophic reactor where algal cells use organic carbon to synthesize oil [144]. Additionally, integration of a two-step strategy with wastewaters, as a nutrient source, can considerably decrease feedstock production costs. Wastewater containing organic compounds generally can be applied as the nutritional sources for heterotrophic or mixotrophic culture media, while wastewaters without organic carbon can be used for phototrophic cultivation mode [145–147].

Up to date, various cultivation modes have been integrated with the treatment of different wastewater types for the production of the microalgae feedstock [142]. Pure simple strategies in phototrophic and heterotrophic, mixotrophic modes, and combinations of phototrophic, heterotrophic, and mixotrophic cultivations on microalgae growth have been studied in several studies (Table 7). These achievements demonstrated the notion of a two-stage strategy as a desirable strategy for obtaining high lipid and biomass concentrations. Xiong et al. [148] compared the efficiency of single-stage heterotrophic mode with photosynthesis–fermentation model on *C. protothecoides*. Photosynthesis–fermentation strategy presented a better performance in terms of lipid productivity (69% higher lipid content) than the one-step heterotrophic mode. Zhou et al. [149] combined a two-step mix-photoautotrophic cultivation strategy with wastewater treatment to produce animal feed and biofuel production. In a similar investigation, a hetero-photoautotrophic microalgal growth mode was investigated for improving wastewater treatment and the production of low-cost algal biofuel feedstock from *Auxenochlorella protothecoides* [150].

Similarly, Liu et al. [151] studied the effects of cultivation strategies including heterotrophic + mixotrophic strategy, heterotrophic strategy, autotrophic cultivation, and heterotrophic + autotrophic strategy on the lipid /biomass production of *Chlorella* sp. HQ. The results showed that the heterotrophic + mixotrophic two-phase culture system was the best strategy for improving the microalgal biomass and heterotrophic cultivation was the best strategy for microalgae lipid accumulation of *Chlorella* sp. HQ. Although several studies are focusing on these aspects, an effective microalgae-based system that does not compromise the growth or lipid content is yet to be found [152]. Some of the obtained lipid/biomass productivity levels of various wastewaters as growth media are presented in Table 7.

Table 7. Comparison of the achieved lipid/biomass productivities of different wastewaters as growth media for cultivating various algal species.

Microalgae Strain	Biomass Productivity $(g\ L^{-1})$	Lipid Productivity $(mg\ L^{-1}\ day^{-1})$	Metabolism Mode/ Type of Wastewater	Refs.
Chlorella vulgaris	4.9	80	Salinity + Nitrogen starvation/wastewater	[140]
Phaeodactylum tricornutum	54.76	54.76	Mixed municipal wastewater and seawater	[153]
Chlorella vulgaris	226.6	108	two-stage photoautotrophic photoheterotrophic/mixotrophic mode	[154]
Chlorella sorokiniana	>7	83.5%	Three-stage cultivation/Farm wastewater	[155]
Chlorella vulgaris	2.92	163	Wastewater + glycerol addition	[156]
Chlorella vulgaris	1.89 ± 0.07	24.7 ± 1.2	Photo-mixotrophic/ Centrate wastewater	[157]

Irradiation

Manipulation of irradiation-based stimuli, in terms of intensity and wavelength, has been recognized as a potential approach for enhancing lipid content in various kinds of cultivation strategies [61,126]. He et al. [22] found that fluctuating the light intensities (990 to 1486 μ mol photons $m^{-2}\ s^{-1}$) increased the lipids yield of six microalgae strains. In another study, increasing the light intensity from 300 to 500 μ mol m^{-2} s^{-1} considerably increased the lipid productivity of *Nannochloropsis oculata*. In this study, the lipid productivity value obtained in the two-step strategy was nearly three times more than the one-step strategy [137].

Apart from light intensity, light frequency is also important in photosynthesis; it can influence algal biomass productivity [45]. Microalgae have a light-harvesting antenna, which primarily absorbs light wavelengths in the visible spectrum [158]. It has been shown that the coupling of red light-emitting diode (LED) and 10–30% blue LED provides the proper light frequency for feedstock production. In contrast, green wavelengths cannot be absorbed by microalgae cells but may have a positive impact on increasing lipid accumulation in microalgae [126,159,160]. Therefore, integrating different wavelengths of lights, with a two-stage strategy can be an efficient strategy in optimization algal biomass and lipid productivities. So that, in the first stage, the microalgae were cultivated under red and/or blue LEDs to achieve maximal biomass productivity. Then, in the second stage, a green LED (520 nm) stress was stimulated to increase lipids accumulation [126]. In an investigation, LEDs were applied to improved cell growth and lipid production of *Picochlorum atomus* by a two-step strategy. The results indicate that biomass productivity under red LED light was higher than that produced by yellow, blue, and purple LEDs in the first phase. The highest lipid production was achieved (50.3%) with green LED light in the second stage [161].

3.3.3. Combined Cultivation Strategies

Although microalgae can produce higher lipids/biomass productivity compared with terrestrial plants, commercialization of the microalgae-derived biofuels is hampered by the production of the high cell density culture with high lipid content [162]. Generally, each of the cultivation strategies has its own merits in terms of the biomass/lipid yield. It might be favorable to integrate two or more cultivation systems to obtain higher lipid/biomass productivity. Notably, whether the combined or single cultivation systems are applied; it is vital to ensure that microalgae-based-biofuel production with the proposed strategies is more feasible than the conventional culture strategy. For instance, the combination of a two-stage system with a fed-batch strategy led to higher lipid productivity in comparison with a fed-batch system [163]. Similarly, the integration of a semi-continuous system with a two-stage strategy could also offer better efficiency in terms of increasing lipid productivity in *Neochloris oleoabundans* [164].

In addition, the analysis of the biological characteristic of various algae species and optimization of cultivation strategies are of paramount importance in improving lipid and biomass productivity. For instance, Nayak et al. [165] optimized a continuous two-step strategy of *Chlorella* sp. HS2, with supplementation of additional phosphorus at the start of nitrogen starvation in the second phase. In other investigations, Ghidossi et al. [166] proposed an efficient two-stage fed-batch cultivation strategy based on the carbon to nitrogen mass ratio (C/N) the culture medium. In the first phase of this strategy, high cell concentrations were used under carbon starvation (lower C/N ratios). In the second phase, high lipid content was obtained under nitrogen depletion conditions (higher C/N ratios). In this study, lipid productivity attained 2- to 5-fold increase compared to other studies [166–168].

4. Downstream Measure

It is documented that determining the strategies that provide the best performance in harvesting, and after-harvesting stages are of paramount importance in achieving the commercially feasible feedstock production process [20]. The downstream technologies generally include the best harvesting and after-harvesting methods without having qualitative damages to algal biomass and/or lipid content [20,50,51].

4.1. Harvesting Stage

Harvesting is an important stage in the feedstock production process that requires to be investigated carefully throughout an integrated approach [49,51]. Accordingly, establishing the strategies that provide the best performance without qualitative damages to FA profiles, biomass, and lipid contents is crucial in implementing successful feedstock production [51]. Some strategies have been developed for algal biomass recovering, which the major ones include membrane filtration, centrifugation, and coagulation [124]. The effects of various harvesting methods on FA change and cell damage are presented in Figure 4.

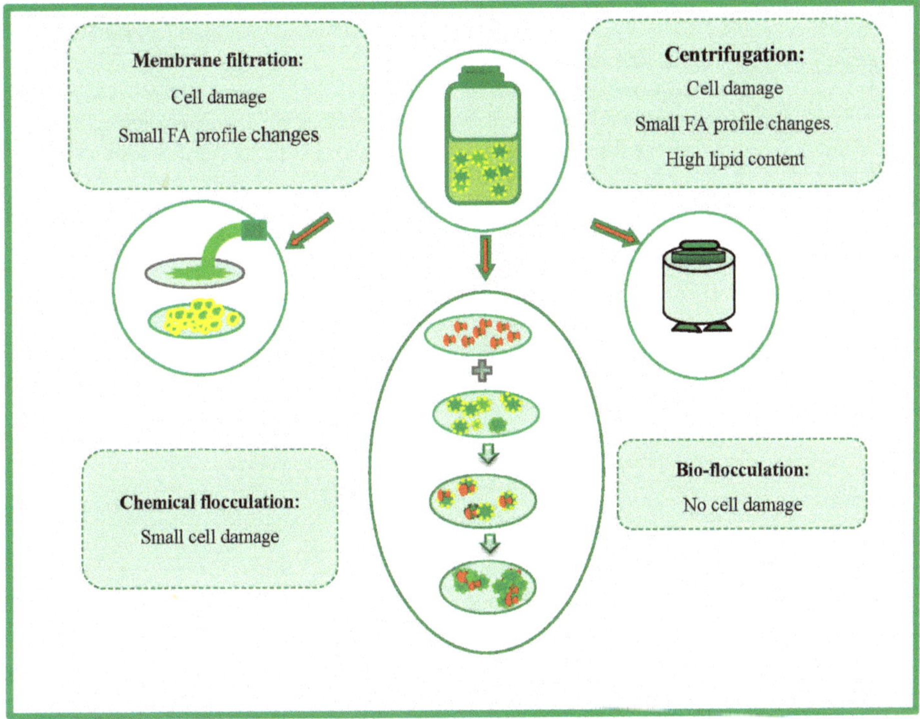

Figure 4. Effects of different harvesting methods on fatty acid (FA) changes and cell damage.

Flocculation is a convenient harvesting strategy for recovering large volumes of algal biomass [169]. It can be performed by conventional harvesting methods, e.g., bioflocculation, chemical flocculation, or novel strategies such as magnetic nanoparticles [170]. It was reported that alum and alkaline flocculation did not severely affect total lipid content. In an investigation, Chatsungnoen et al. [171] found that metal salts such as ferric chloride and aluminum sulfate irreversibly bind to the biomass of *Neochloris*, *Nannochloropsis*, and *Chlorella*, sp. However, these metal coagulants did not significantly impact the biomass and total lipid content. In another study, Vandamme et al. [51] obtained slightly lower lipid content by alkaline or alum flocculation. The reduced lipid content might be explainable by the algal extracellular organic matter (EOM) interaction with the used coagulants [51,172]. Bioflocculants can reduce the demand for chemical flocculants. However, the competition between the bacteria and algae is a drawback for co-cultivation of bioflocculant-producing bacteria and microalgae, which subsequently can affect algal lipid content and cell density. In the recent past, some harvesting strategies have been proposed as promising approaches for algal biomass recovering. Among them, electro-coagulation–floatation (ECF) is considered a good substitute to conventional harvesting methods such as the chemical flocculation approach, due to the low energy requirement and no direct use of coagulant [173]. Fayad et al. [174] applied ECF, using iron and aluminum electrodes, for the harvesting of *Chlorella vulgaris* biomass and reported this harvesting method to have no impact on the amount of algal lipid production.

Additionally, magnetic particles have been described as an interesting flocculation option for microalgal harvesting, in which suspended algae cells were adsorbed or tagged to nano-sized or micron-sized magnetic particles. The tagged composites were separated using external magnet force because of intrinsic paramagnetic movement [175]. Liu et al. [52] applied magnetite nanoparticles (nano-Fe_3O_4 covered with polyethyleneimine) to recover microalgae *C. pyrenoidosa* and *S. obliquus*, and reported the process did not reduce the lipid content of microalgae. In another study, *C. pyrenoidosa* cells were harvested by the method of flocculation using Fe_3O_4-silica nanoparticles for improving microalgae lipid production [176].

As for the FA profile, contradicting results have been reported in different harvesting methods. Some studies showed significant differences in obtained FA profiles in various recovering strategies, but in other investigations, the differences were not significant (Table 8). For instance, in a study, three harvesting methods of centrifugation, microfiltration membrane, and coagulation were analyzed in *Chlorella* sp. biomass recovering. Coagulation was found to exhibit the most poorly results in terms of obtained FA profiles compared to centrifugation and microfiltration membranes [177,178].

Membrane filtration is another efficient harvesting strategy for the aggregation of microalgal cells. Operating under a low trans-membrane pressure makes the approach less energy-intensive than centrifugation, and the long membrane lifespan makes the harvesting process more cost-effective in the long term [179]. Membrane-based separation processes also pose several challenges such as membrane clogging, and electrostatic repulsion from the negative surface charges of microalgae cells and membrane surface.

Table 8. Lipid recovery under various harvesting strategies.

Microalga Species	Harvesting Method	FAME Profile	Lipid Yield (%)	Ref.
Chlorella sp.	flotation + centrifugation	MUFA increase	13.4	[180]
	Centrifugation	PUFA increase	9.9	
Chlorella sp.	Centrifugation	No change	27	[178]
	Microfiltration	No change	26	
	Coagulation	No change	15	
Nannochloropsis oculata	Flocculation with NaOH	The balance between SFA, PUFA, and MUFA	4.3	[177]
	Flocculation with Magnafloc	Lower C20:5 and higher C14:0 and	~4.4	
	Filtration	The balance between, SFA, PUFA, and MUFA	3.6	
Thalassiosira weissflogi	Flocculation with Flopam	C18:0, C18:1n9c, and C16 0, increase	4.12	
		The balance between, SFA, PUFA, and MUFA	~3.1	
		The balance between MUFA, PUFA, and SFA	2.77	

FAME: FA methyl esters; SFA: Saturated FAs; MUFA: Monounsaturated FA (%); PUFA: Polyunsaturated FAs.

Centrifugation represents an alternative strategy that can be used in pilot-scale production. The harvesting by centrifugation generally presents better results in terms of lipid content when compared to filtration or flocculation [178,179]. However, it should be noted that algal cells exposed to high gravitational forces during centrifugation can result in structural cell damage and small FA profile changes. The effects of harvesting strategy on biomass quality are important when biochemical components must meet quality standards for further processing of obtained algal biomass, e.g., lipids for biodiesel production (Figure 3) [49].

4.2. Post-Harvesting Stage

Lipid enhancement approaches involving alteration of environmental conditions and nutrient limitation regimes such as temperature, light, and nutrient limitation (e.g., phosphorus and nitrogen) are conventional strategies applied to increase lipid production in microalgae [16]. Generally, conventional stress-inducing methods are applied during the cultivation stage [19]. However, recent investigations have shown that the application of different strategies in other stages of the feedstock production process also can improve lipid and/or biomass production in microalgae [51,123,181]. It was reported that the stress induction in the post-harvesting stage resulted in a positive effect in the lipid accumulation of *Chlorella vulgaris* [50]. In this investigation, the high lipid content in the after-harvesting stage was achieved in one day of nutrients starvation under dark conditions. The lipid production under these conditions was considerably increased compared to the control group. In addition, it was demonstrated that the main produced FAs of *C. vulgaris* were oleic acid, palmitic acid, and linoleic acid, sharing similar FA profiles to those of soybean, sunflower, and corn oil [50].

Additionally, microalgae-based biofuels are produced from algal feedstock by thermochemical, biochemical, and chemical methods. Among the thermochemical techniques, pyrolysis is considered as a potential approach involving high pressure and high temperature to produce bio-oil and biochar from the microalgal feedstock. Therefore, selecting a suitable pyrolysis method can influence obtaining a desirable quality and quantity of bio-oil from algae [182].

5. Conclusions

Microalgae have received widespread interest owing to their unique properties in producing large amounts of lipids and fast growth rates. However, algal strains exhibit conflicting features in terms of the conditions required for maximal lipid and biomass production. These contradictory features can be mitigated by applying appropriate strategies throughout the biomass production process. The purpose of this article is to review the technologies and advancements available for enhancing lipid productivity in microalgae species. The first step in the feedstock production process is screening the right alga with relevant properties and further improvement of those platform species by genetic manipulation. Approaches like genetic modification at the metabolic and genomic levels can be beneficial in improving biomass/lipid production. Additionally, it has been reported that, throughout cultivation, threshold nutrients give lower lipid content but high biomass yield and vice versa. These conditions can be mitigated by optimizing an appropriate two-phase cultivation strategy. Apart from the two-stage cultivation strategy, alteration growth parameters in some other cultivation strategies such as continuous chemostat can also produce higher lipid/biomass yield. Establishing the strategies that provide the best performance in the harvesting stage without qualitative damages to microalgae biomass and/or lipid content is important in the successful implementation of the feedstock production process. Magnetic and Bio-based flocculants are very promising for algal biomass recovering. Thus, choosing an appropriate screening, cultivation, harvesting, and after-harvesting strategies can influence obtaining desirable quantity and quality of bio-oil from algae. It is our hope that this review article could inspire ongoing efforts in developing sustainable microalgae-derived biofuel production with improved biomass/lipid yield at an economically feasible cost.

Author Contributions: Conceptualization, Z.S., and H.S.; software, Z.S., H.S., O.H.C., S.S.R.K., and M.P.; validation, Z.S.; formal analysis, Z.S., and H.S.; investigation, Z.S.; resources, Z.S., H.S., O.H.C., W.J.L., S.S.R.K., M.P., and A.F.I.; data curation, Z.S.; writing—original draft preparation, Z.S.; writing—review and editing, Z.S., H.S., O.H.C., W.J.L., and A.F.I.; visualization, Z.S., H.S., O.H.C., W.J.L., S.S.R.K., M.P., and A.F.I.; project administration, Z.S., H.S., O.H.C., W.J.L., S.S.R.K., M.P., and A.F.I.; funding acquisition, Z.S., H.S., O.H.C., W.J.L., S.S.R.K., M.P., and A.F.I. All authors have read and agreed to the published version of the manuscript.

Funding: Ministry of Education, Youth, and Sports of the Czech Republic and the European Union (European Structural and Investment Funds Operational Program Research, Development, and Education) in the framework of the project "Modular platform for autonomous chassis of specialized electric vehicles for freight and equipment transportation", Reg. No. CZ.02.1.01/0.0/0.0/16_025/0007293, as well as the financial support from internal grants in the Institute for Nanomaterials, Advanced Technologies and Innovations (CXI), Technical University of Liberec (TUL).

Conflicts of Interest: The authors declare no conflict of interest.

References

1. Singh, D.; Sharma, D.; Soni, S.; Sharma, S.; Sharma, P.K.; Jhalani, A. A review on feedstocks, production processes, and yield for different generations of biodiesel. *Fuel* **2020**, *262*, 116553. [CrossRef]

2. Günay, M.E.; Türker, L.; Tapan, N.A. Significant parameters and technological advancements in biodiesel production systems. *Fuel* **2019**, *250*, 27–41. [CrossRef]

3. Milano, J.; Ong, H.C.; Masjuki, H.; Chong, W.; Lam, M.K.; Loh, P.K.; Vellayan, V. Microalgae biofuels as an alternative to fossil fuel for power generation. *Renew. Sustain. Energy Rev.* **2016**, *58*, 180–197. [CrossRef]

4. Ullah, K.; Ahmad, M.; Sharma, V.K.; Lu, P.; Harvey, A.; Zafar, M.; Sultana, S. Assessing the potential of algal biomass opportunities for bioenergy industry: A review. *Fuel* **2015**, *143*, 414–423. [CrossRef]

5. Shanmugam, S.; Hari, A.; Pandey, A.; Mathimani, T.; Felix, L.; Pugazhendhi, A. Comprehensive review on the application of inorganic and organic nanoparticles for enhancing biohydrogen production. *Fuel* **2020**, *270*, 117453. [CrossRef]

6. Sharma, Y.; Singh, B.; Upadhyay, S. Advancements in development and characterization of biodiesel: A review. *Fuel* **2008**, *87*, 2355–2373. [CrossRef]

7. Stephen, J.L.; Periyasamy, B. Innovative developments in biofuels production from organic waste materials: A review. *Fuel* **2018**, *214*, 623–633. [CrossRef]

8. Vassilev, S.V.; Vassileva, C.G. Composition, properties and challenges of algae biomass for biofuel application: An overview. *Fuel* **2016**, *181*, 1–33. [CrossRef]

9. Khalid, A.A.H.; Yaakob, Z.; Abdullah, S.R.S.; Takriff, M.S. Growth improvement and metabolic profiling of native and commercial Chlorella sorokiniana strains acclimatized in recycled agricultural wastewater. *Bioresour. Technol.* **2018**, *247*, 930–939. [CrossRef]

10. Deng, X.; Li, Y.; Fei, X. Microalgae: A promising feedstock for biodiesel. *Afr. J. Microbiol. Res.* **2009**, *3*, 1008–1014.

11. Nayak, M.; Suh, W.I.; Chang, Y.K.; Lee, B. Exploration of two-stage cultivation strategies using nitrogen starvation to maximize the lipid productivity in *Chlorella* sp. HS2. *Bioresour. Technol.* **2019**, *276*, 110–118. [CrossRef] [PubMed]

12. Shokravi, H.; Shokravi, Z.; Heidarrezaei, M.; Lau, W.J.; Koloor, S.S.R.; Petrů, M.; Chyuan, O.H.; Ismail, A.F. Fourth Generation Biofuel: A Review on Challenges and Future Directions. *Fuel* **2020**, under review.

13. Adams, C.; Godfrey, V.; Wahlen, B.; Seefeldt, L.; Bugbee, B. Understanding precision nitrogen stress to optimize the growth and lipid content tradeoff in oleaginous green microalgae. *Bioresour. Technol.* **2013**, *131*, 188–194. [CrossRef] [PubMed]

14. Chen, B.; Wan, C.; Mehmood, M.A.; Chang, J.-S.; Bai, F.; Zhao, X. Manipulating environmental stresses and stress tolerance of microalgae for enhanced production of lipids and value-added products—A review. *Bioresour. Technol.* **2017**, *244*, 1198–1206. [CrossRef]

15. Ghosh, A.; Khanra, S.; Mondal, M.; Halder, G.; Tiwari, O.; Saini, S.; Bhowmick, T.K.; Gayen, K. Progress toward isolation of strains and genetically engineered strains of microalgae for production of biofuel and other value added chemicals: A review. *Energy Convers. Manag.* **2016**, *113*, 104–118. [CrossRef]

16. Sajjadi, B.; Chen, W.-Y.; Raman, A.A.A.; Ibrahim, S. Microalgae lipid and biomass for biofuel production: A comprehensive review on lipid enhancement strategies and their effects on fatty acid composition. *Renew. Sustain. Energy Rev.* **2018**, *97*, 200–232. [CrossRef]

17. Shokravi, H.; Shokravi, Z.; Aziz, M.A.; Shokravi, H. 11 Algal Biofuel: A Promising. In *Fossil Free Fuels: Trends in Renewable Energy*; Taylor & Francis Group: Oxford, UK, 2019; pp. 187–211.

18. Shokravi, H.; Shokravi, Z.; Aziz, M.A.; Shokravi, H. 12 The Fourth-Generation. In *Fossil Free Fuels: Trends in Renewable Energy*; Taylor & Francis Group: Oxford, UK, 2019; pp. 213–251.

19. Ho, S.-H.; Ye, X.; Hasunuma, T.; Chang, J.-S.; Kondo, A. Perspectives on engineering strategies for improving biofuel production from microalgae—A critical review. *Biotechnol. Adv.* **2014**, *32*, 1448–1459. [CrossRef]

20. Peng, L.; Fu, D.; Chu, H.; Wang, Z.; Qi, H. Biofuel production from microalgae: A review. *Environ. Chem. Lett.* **2019**, *18*, 285–297. [CrossRef]

21. Lim, D.K.Y.; Schenk, P.M. Microalgae selection and improvement as oil crops: GM vs non-GM strain engineering. *AIMS Environ. Sci.* **2017**, *4*, 151–161. [CrossRef]

22. He, Q.; Yang, H.; Xu, L.; Xia, L.; Hu, C. Sufficient utilization of natural fluctuating light intensity is an effective approach of promoting lipid productivity in oleaginous microalgal cultivation outdoors. *Bioresour. Technol.* **2015**, *180*, 79–87. [CrossRef]

23. Přibyl, P.; Cepák, V.; Zachleder, V. Production of lipids in 10 strains of Chlorella and Parachlorella, and enhanced lipid productivity in *Chlorella vulgaris*. *Appl. Microbiol. Biotechnol.* **2012**, *94*, 549–561. [CrossRef]

24. Kwak, H.S.; Kim, J.Y.H.; Woo, H.M.; Jin, E.; Min, B.K.; Sim, S.J. Synergistic effect of multiple stress conditions for improving microalgal lipid production. *Algal Res.* **2016**, *19*, 215–224. [CrossRef]

25. Kim, H.S.; Hsu, S.-C.; Han, S.-I.; Thapa, H.R.; Guzman, A.R.; Browne, D.R.; Tatli, M.; Devarenne, T.P.; Stern, D.B.; Han, A. High-throughput droplet microfluidics screening platform for selecting fast-growing and high lipid-producing microalgae from a mutant library. *Plant Direct* **2017**, *1*, e00011. [CrossRef]

26. Challagulla, V.; Nayar, S.; Walsh, K.; Fabbro, L. Advances in techniques for assessment of microalgal lipids. *Crit. Rev. Biotechnol.* **2016**, *37*, 566–578. [CrossRef]

27. Cabanelas, I.T.D.; Van Der Zwart, M.; Kleinegris, D.M.M.; Barbosa, M.J.; Wijffels, R.H. Rapid method to screen and sort lipid accumulating microalgae. *Bioresour. Technol.* **2015**, *184*, 47–52. [CrossRef]

28. Erickson, R.A.; Jimenez, R. Microfluidic cytometer for high-throughput measurement of photosynthetic characteristics and lipid accumulation in individual algal cells. *Lab Chip* **2013**, *13*, 2893–2901. [CrossRef]

29. Liu, J.; Mukherjee, J.; Hawkes, J.J.; Wilkinson, S.J. Optimization of lipid production for algal biodiesel in nitrogen stressed cells of *Dunaliella salina* using FTIR analysis. *J. Chem. Technol. Biotechnol.* **2013**, *88*, 1807–1814. [CrossRef]

30. Shokravi, Z.; Mehrad, L.; Ramazani, A. Detecting the frequency of aminoglycoside modifying enzyme encoding genes among clinical isolates of methicillin-resistant *Staphylococcus aureus*. *BioImpacts* **2015**, *5*, 87–91. [CrossRef]

31. Shokravi, Z.; Haseli, M.; Mehrad, L.; Ramazani, A. Distribution of Staphylococcal cassette chromosome mecA (SCCmec) types among coagulase-negative Staphylococci isolates from healthcare workers in the North-West of Iran. *Iran. J. Basic Med. Sci.* **2020**, *23*, 1489–1493. [CrossRef]

32. Ramazani, A.; Shokravi, Z.; Mehrad, L.; Sorouri, R.; Amirmoghaddami, H. Study the Genes Encoding Aminoglycoside-Modifying Enzymes among Clinical Isolates of Methicillin Resistance Staphylococcus Aureus. 2013. Available online: https://www.sid.ir/en/Seminar/ViewPaper.aspx?ID=32537 (accessed on 29 October 2020).

33. Heidarrezaei, M.; Shokravi, H.; Huyop, F.; Koloor, S.R.; Petru, M. Isolation and Characterization of a Novel Bacterium from the Marine Environment for Trichloroacetic Acid Bioremediation. *Appl. Sci.* **2020**, *10*, 4593. [CrossRef]

34. Sharma, P.K.; Saharia, M.; Srivetava, R.; Kumar, S.; Sahoo, L. Tailoring Microalgae for Efficient Biofuel Production. *Front. Mar. Sci.* **2018**, *5*, 382. [CrossRef]

35. Trentacoste, E.M.; Shrestha, R.P.; Smith, S.R.; Glé, C.; Hartmann, A.C.; Hildebrand, M.; Gerwick, W.H. Metabolic engineering of lipid catabolism increases microalgal lipid accumulation without compromising growth. *Proc. Natl. Acad. Sci. USA* **2013**, *110*, 19748–19753. [CrossRef]

36. Nguyen, T.H.T.; Park, S.; Jeong, J.; Shin, Y.S.; Sim, S.J.; Jin, E. Enhancing lipid productivity by modulating lipid catabolism using the CRISPR-Cas9 system in Chlamydomonas. *Environ. Biol. Fishes* **2020**, *32*, 2829–2840. [CrossRef]

37. Park, S.; Nguyen, T.H.T.; Jin, E. Improving lipid production by strain development in microalgae: Strategies, challenges and perspectives. *Bioresour. Technol.* **2019**, *292*, 121953. [CrossRef]

38. Xia, L.; Ge, H.; Zhou, X.; Zhang, D.; Hu, C. Photoautotrophic outdoor two-stage cultivation for oleaginous microalgae Scenedesmus obtusus XJ-15. *Bioresour. Technol.* **2013**, *144*, 261–267. [CrossRef] [PubMed]

39. Yun, J.-H.; Cho, D.-H.; Lee, S.; Heo, J.; Tran, Q.-G.; Chang, Y.K.; Kim, H.-S. Hybrid operation of photobioreactor and wastewater-fed open raceway ponds enhances the dominance of target algal species and algal biomass production. *Algal Res.* **2018**, *29*, 319–329. [CrossRef]

40. Nagappan, S.; Devendran, S.; Tsai, P.-C.; Dahms, H.-U.; Ponnusamy, V.K. Potential of two-stage cultivation in microalgae biofuel production. *Fuel* **2019**, *252*, 339–349. [CrossRef]

41. Aziz, M.A.; Kassim, K.A.; Shokravi, Z.; Jakarni, F.M.; Liu, H.Y.; Zaini, N.; Tan, L.S.; Islam, A.S.; Shokravi, H.; Lieu, H.Y. Two-stage cultivation strategy for simultaneous increases in growth rate and lipid content of microalgae: A review. *Renew. Sustain. Energy Rev.* **2020**, *119*, 109621. [CrossRef]

42. Bhatia, S.; Mehariya, S.; Bhatia, R.K.; Kumar, M.; Pugazhendhi, A.; Awasthi, M.K.; Atabani, A.; Kumar, G.; Kim, W.; Seo, S.-O.; et al. Wastewater based microalgal biorefinery for bioenergy production: Progress and challenges. *Sci. Total. Environ.* **2021**, *751*, 141599. [CrossRef]

43. Borowitzka, M.A. Species and strain selection. In *Algae for Biofuels and Energy*; Springer: Berlin, Germany, 2013; pp. 77–89.

44. Singh, P.; Kumari, S.; Guldhe, A.; Misra, R.; Rawat, I.; Bux, F. Trends and novel strategies for enhancing lipid accumulation and quality in microalgae. *Renew. Sustain. Energy Rev.* **2016**, *55*, 1–16. [CrossRef]

45. Chu, W.-L. Strategies to enhance production of microalgal biomass and lipids for biofuel feedstock. *Eur. J. Phycol.* **2017**, *52*, 419–437. [CrossRef]

46. Chung, Y.-S.; Lee, J.-W.; Chung, C.-H. Molecular challenges in microalgae towards cost-effective production of quality biodiesel. *Renew. Sustain. Energy Rev.* **2017**, *74*, 139–144. [CrossRef]

47. Shin, Y.S.; Choi, H.I.; Choi, J.W.; Lee, J.S.; Sung, Y.J.; Sim, S.J. Multilateral approach on enhancing economic viability of lipid production from microalgae: A review. *Bioresour. Technol.* **2018**, *258*, 335–344. [CrossRef] [PubMed]

48. Piligaev, A.V.; Sorokina, K.N.; Samoylova, Y.V.; Parmon, V.N. Production of Microalgal Biomass with High Lipid Content and Their Catalytic Processing Into Biodiesel: A Review. *Catal. Ind.* **2019**, *11*, 349–359. [CrossRef]

49. Menegazzo, M.L.; Fonseca, G.G. Biomass recovery and lipid extraction processes for microalgae biofuels production: A review. *Renew. Sustain. Energy Rev.* **2019**, *107*, 87–107. [CrossRef]

50. Poh, Z.L.; Kadir, W.N.A.; Lam, M.-K.; Uemura, Y.; Suparmaniam, U.; Lim, J.W.; Show, P.L.; Lee, K.T. The effect of stress environment towards lipid accumulation in microalgae after harvesting. *Renew. Energy* **2020**, *154*, 1083–1091. [CrossRef]

51. Vandamme, D.; Gheysen, L.; Muylaert, K.; Foubert, I. Impact of harvesting method on total lipid content and extraction efficiency for Phaeodactylum tricornutum. *Sep. Purif. Technol.* **2018**, *194*, 362–367. [CrossRef]

52. Liu, Y.; Jin, W.; Zhou, X.; Han, S.-F.; Tu, R.; Feng, X.; Jensen, P.D.; Wang, Q. Efficient harvesting of Chlorella pyrenoidosa and Scenedesmus obliquus cultivated in urban sewage by magnetic flocculation using nano-Fe3O4 coated with polyethyleneimine. *Bioresour. Technol.* **2019**, *290*, 121771. [CrossRef]

53. Nwokoagbara, E.; Olaleye, A.K.; Wang, M. Biodiesel from microalgae: The use of multi-criteria decision analysis for strain selection. *Fuel* **2015**, *159*, 241–249. [CrossRef]

54. Sibi, G.; Shetty, V.; Mokashi, K. Enhanced lipid productivity approaches in microalgae as an alternate for fossil fuels—A review. *J. Energy Inst.* **2016**, *89*, 330–334. [CrossRef]

55. Gao, Y.; Yang, M.; Wang, C. Nutrient deprivation enhances lipid content in marine microalgae. *Bioresour. Technol.* **2013**, *147*, 484–491. [CrossRef] [PubMed]

56. Li, T.; Wan, L.; Li, A.; Zhang, C. Responses in growth, lipid accumulation, and fatty acid composition of four oleaginous microalgae to different nitrogen sources and concentrations. *Chin. J. Oceanol. Limnol.* **2013**, *31*, 1306–1314. [CrossRef]

57. Yu, N.; Dieu, L.T.J.; Harvey, S.; Lee, D.-Y. Optimization of process configuration and strain selection for microalgae-based biodiesel production. *Bioresour. Technol.* **2015**, *193*, 25–34. [CrossRef] [PubMed]

58. Sun, H.; Ren, Y.; Mao, X.; Li, X.; Zhang, H.; Lao, Y.; Chen, F. Harnessing C/N balance of *Chromochloris zofingiensis* to overcome the potential conflict in microalgal production. *Commun. Biol.* **2020**, *3*, 1–13. [CrossRef]

59. Hsieh, H.-J.; Su, C.-H.; Chien, L.-J. Accumulation of lipid production in *Chlorella minutissima* by triacylglycerol biosynthesis-related genes cloned from *Saccharomyces cerevisiae* and *Yarrowia lipolytica*. *J. Microbiol.* **2012**, *50*, 526–534. [CrossRef]

60. Moussa, I.D.-B.; Chtourou, H.; Hassairi, I.; Sayadi, S.; Dhouib, A. The effect of switching environmental conditions on content and structure of lipid produced by a wild strain *Picochlorum* sp. *Renew. Energy* **2019**, *134*, 406–415. [CrossRef]

61. Pedro, A.S.; González-López, C.; Acién, F.; Molina-Grima, E. Marine microalgae selection and culture conditions optimization for biodiesel production. *Bioresour. Technol.* **2013**, *134*, 353–361. [CrossRef]

62. Pal, P.; Chew, K.W.; Yen, H.-W.; Lim, J.W.; Lam, M.K.; Show, P.L. Cultivation of oily microalgae for the production of third-generation biofuels. *Sustainability* **2019**, *11*, 5424. [CrossRef]

63. De Bhowmick, G.; Koduru, L.; Sen, R. Metabolic pathway engineering towards enhancing microalgal lipid biosynthesis for biofuel application—A review. *Renew. Sustain. Energy Rev.* **2015**, *50*, 1239–1253. [CrossRef]

64. El Arroussi, H.; Benhima, R.; El Mernissi, N.; Bouhfid, R.; Tilsaghani, C.; Bennis, I.; Wahby, I. Screening of marine microalgae strains from Moroccan coasts for biodiesel production. *Renew. Energy* **2017**, *113*, 1515–1522. [CrossRef]

65. Huang, S.T.; Goh, J.L.; Ahmadzadeh, H.; Murry, M.A. A rapid sampling technique for isolating highly productive lipid-rich algae strains from environmental samples. *Biofuel Res. J.* **2019**, *6*, 920–926. [CrossRef]

66. Neofotis, P.; Huang, A.; Sury, K.; Chang, W.; Joseph, F.; Gabr, A.; Twary, S.; Qiu, W.; Holguin, O.; Polle, J.E. Characterization and classification of highly productive microalgae strains discovered for biofuel and bioproduct generation. *Algal Res.* **2016**, *15*, 164–178. [CrossRef]

67. Andersen, R.A.; Kawachi, M. Traditional Microalgae Isolation Techniques. In *Algal Culturing Techniques*; Elsevier Academic Press: Burlington, MA, USA, 2005; pp. 83–100.

68. Mutanda, T.; Ramesh, D.; Karthikeyan, S.; Kumari, S.; Anandraj, A.; Bux, F. Bioprospecting for hyper-lipid producing microalgal strains for sustainable biofuel production. *Bioresour. Technol.* **2011**, *102*, 57–70. [CrossRef] [PubMed]

69. Cabanelas, I.T.D.; Van Der Zwart, M.; Kleinegris, D.M.M.; Wijffels, R.H.; Barbosa, M.J. Sorting cells of the microalga *Chlorococcum littorale* with increased triacylglycerol productivity. *Biotechnol. Biofuels* **2016**, *9*, 1–12. [CrossRef]

70. Pereira, H.; Barreira, L.; Mozes, A.; Florindo, C.; Polo, C.; Duarte, C.V.; Custodio, L.; Varela, J. Microplate-based high throughput screening procedure for the isolation of lipid-rich marine microalgae. *Biotechnol. Biofuels* **2011**, *4*, 61. [CrossRef] [PubMed]

71. Kim, H.S.; Guzman, A.R.; Thapa, H.R.; Devarenne, T.P.; Han, A. A droplet microfluidics platform for rapid microalgal growth and oil production analysis. *Biotechnol. Bioeng.* **2016**, *113*, 1691–1701. [CrossRef]

72. Del Río, E.; Acién, F.G.; Guerrero, M.G.; Sanchez, E.D.R. Photoautotrophic production of astaxanthin by the microalga *Haematococcus pluvialis*. In *Sustainable Biotechnology*; Springer: Dordrecht, The Netherlands, 2010; pp. 247–258.

73. Sung, M.-G.; Lee, B.; Kim, C.W.; Nam, K.; Chang, Y.K. Enhancement of lipid productivity by adopting multi-stage continuous cultivation strategy in *Nannochloropsis gaditana*. *Bioresour. Technol.* **2017**, *229*, 20–25. [CrossRef]

74. Abdullah, B.; Muhammad, S.A.F.S.; Shokravi, Z.; Ismail, S.; Kassim, K.A.; Mahmood, A.N.; Aziz, M.A. Fourth generation biofuel: A review on risks and mitigation strategies. *Renew. Sustain. Energy Rev.* **2019**, *107*, 37–50. [CrossRef]

75. Davis, R.; Aden, A.; Pienkos, P.T. Techno-economic analysis of autotrophic microalgae for fuel production. *Appl. Energy* **2011**, *88*, 3524–3531. [CrossRef]

76. Eguiheneuf, F.; Ekhan, A.; Tran, L.-S.P. Genetic Engineering: A Promising Tool to Engender Physiological, Biochemical, and Molecular Stress Resilience in Green Microalgae. *Front. Plant Sci.* **2016**, *7*, 400. [CrossRef]

77. Tan, K.W.M.; Lee, Y.-K. The dilemma for lipid productivity in green microalgae: Importance of substrate provision in improving oil yield without sacrificing growth. *Biotechnol. Biofuels* **2016**, *9*, 255. [CrossRef] [PubMed]

78. Talebi, A.F.; Tohidfar, M.; Bagheri, A.; Lyon, S.R.; Salehi-Ashtiani, K.; Tabatabaei, M. Manipulation of carbon flux into fatty acid biosynthesis pathway in *Dunaliella salina* using AccD and ME genes to enhance lipid content and to improve produced biodiesel quality. *Biofuel Res. J.* **2014**, *1*, 91–97. [CrossRef]

79. Xue, J.; Niu, Y.-F.; Huang, T.; Yang, W.-D.; Liu, J.-S.; Li, H.-Y. Genetic improvement of the microalga *Phaeodactylum tricornutum* for boosting neutral lipid accumulation. *Metab. Eng.* **2015**, *27*, 1–9. [CrossRef]

80. Ma, Y.-H.; Wang, X.; Niu, Y.-F.; Yang, Z.-K.; Zhang, M.-H.; Wang, Z.; Yang, W.-D.; Liu, J.-S.; Li, H.-Y. Antisense knockdown of pyruvate dehydrogenase kinase promotes the neutral lipid accumulation in the diatom *Phaeodactylum tricornutum*. *Microb. Cell Factories* **2014**, *13*, 100. [CrossRef]

81. Muto, M.; Tanaka, M.; Liang, Y.; Yoshino, T.; Matsumoto, M.; Tanaka, T. Enhancement of glycerol metabolism in the oleaginous marine diatom *Fistulifera solaris* JPCC DA0580 to improve triacylglycerol productivity. *Biotechnol. Biofuels* **2015**, *8*, 4. [CrossRef] [PubMed]

82. Fan, J.; Ning, K.; Zeng, X.; Luo, Y.; Wang, D.; Hu, J.; Li, J.; Xu, H.; Huang, J.; Wan, M.; et al. Genomic foundation of starch-to-lipid switch in Oleaginous *Chlorella* spp. *Plant Physiol.* **2015**, *169*, 2444–2461.

83. Deng, X.; Li, Y.; Fei, X. The mRNA abundance of pepc2 gene is negatively correlated with oil content in *Chlamydomonas reinhardtii*. *Biomass Bioenergy* **2011**, *35*, 1811–1817. [CrossRef]

84. Muñoz, C.F.; Weusthuis, R.A.; D'Adamo, S.; Wijffels, R.H. Effect of Single and Combined Expression of Lysophosphatidic Acid Acyltransferase, Glycerol-3-Phosphate Acyltransferase, and Diacylglycerol Acyltransferase on Lipid Accumulation and Composition in *Neochloris oleoabundans*. *Front. Plant Sci.* **2019**, *10*, 1573. [CrossRef]

85. Shahid, A.; Rehman, A.U.; Usman, M.; Ashraf, M.U.F.; Javed, M.R.; Khan, A.Z.; Gill, S.S.; Mehmood, M.A. Engineering the metabolic pathways of lipid biosynthesis to develop robust microalgal strains for biodiesel production. *Biotechnol. Appl. Biochem.* **2020**, *67*, 41–51. [CrossRef]

86. Ran, W.; Wang, H.; Liu, Y.; Qi, M.; Xiang, Q.; Yao, C.; Zhang, Y.; Lan, X. Storage of starch and lipids in microalgae: Biosynthesis and manipulation by nutrients. *Bioresour. Technol.* **2019**, *291*, 121894. [CrossRef]

87. Zhu, B.-H.; Shi, H.-P.; Yang, G.; Lv, N.-N.; Yang, M.; Pan, K.-H. Silencing UDP-glucose pyrophosphorylase gene in *Phaeodactylum tricornutum* affects carbon allocation. *New Biotechnol.* **2016**, *33*, 237–244. [CrossRef]

88. Wan, M.; Jin, X.; Xia, J.; Rosenberg, J.N.; Yu, G.; Nie, Z.; Oyler, G.A.; Betenbaugh, M. The effect of iron on growth, lipid accumulation, and gene expression profile of the freshwater microalga *Chlorella sorokiniana*. *Appl. Microbiol. Biotechnol.* **2014**, *98*, 9473–9481. [CrossRef]

89. Chauton, M.S.; Winge, P.; Brembu, T.; Vadstein, O.; Bones, A.M. Gene Regulation of Carbon Fixation, Storage, and Utilization in the Diatom *Phaeodactylum tricornutum* Acclimated to Light/Dark Cycles. *Plant Physiol.* **2012**, *161*, 1034–1048. [CrossRef]

90. Solovchenko, A. Physiological role of neutral lipid accumulation in eukaryotic microalgae under stresses. *Russ. J. Plant Physiol.* **2012**, *59*, 167–176. [CrossRef]

91. Iwai, M.; Ikeda, K.; Shimojima, M.; Ohta, H. Enhancement of extraplastidic oil synthesis in C hlamydomonas reinhardtii using a type-2 diacylglycerol acyltransferase with a phosphorus starvation–inducible promoter. *Plant Biotechnol. J.* **2014**, *12*, 808–819. [CrossRef] [PubMed]

92. Eiwai, M.; Ehori, K.; Esasaki-Sekimoto, Y.; Eshimojima, M.; Ohta, H. Manipulation of oil synthesis in Nannochloropsis strain NIES-2145 with a phosphorus starvation–inducible promoter from *Chlamydomonas reinhardtii*. *Front. Microbiol.* **2015**, *6*, 912. [CrossRef]

93. Zhang, J.; Hao, Q.; Bai, L.; Xu, J.; Yin, W.-B.; Song, L.; Xu, L.; Guo, X.; Fan, C.; Chen, Y.; et al. Overexpression of the soybean transcription factor GmDof4 significantly enhances the lipid content of *Chlorella ellipsoidea*. *Biotechnol. Biofuels* **2014**, *7*, 1–16. [CrossRef] [PubMed]

94. Tsai, C.-H.; Warakanont, J.; Takeuchi, T.; Sears, B.B.; Moellering, E.R.; Benning, C. The protein Compromised Hydrolysis of Triacylglycerols 7 (CHT7) acts as a repressor of cellular quiescence in Chlamydomonas. *Proc. Natl. Acad. Sci. USA* **2014**, *111*, 15833–15838. [CrossRef]

95. Ngan, C.Y.; Wong, C.-H.; Choi, C.; Yoshinaga, Y.; Louie, K.; Jia, J.; Chen, C.; Bowen, B.; Cheng, H.; Leonelli, L.; et al. Lineage-specific chromatin signatures reveal a regulator of lipid metabolism in microalgae. *Nat. Plants* **2015**, *1*, 15107. [CrossRef]

96. Ajjawi, I.; Verruto, J.; Aqui, M.; Soriaga, L.B.; Coppersmith, J.; Kwok, K.; Peach, L.; Orchard, E.; Kalb, R.; Xu, W.; et al. Lipid production in *Nannochloropsis gaditana* is doubled by decreasing expression of a single transcriptional regulator. *Nat. Biotechnol.* **2017**, *35*, 647–652. [CrossRef] [PubMed]

97. Kang, N.K.; Jeon, S.; Kwon, S.; Koh, H.G.; Shin, S.-E.; Lee, B.; Choi, G.-G.; Yang, J.-W.; Jeong, B.-R.; Chang, Y.K. Effects of overexpression of a bHLH transcription factor on biomass and lipid production in *Nannochloropsis salina*. *Biotechnol. Biofuels* **2015**, *8*, 1–13. [CrossRef] [PubMed]

98. Kwon, S.; Kang, N.K.; Koh, H.G.; Shin, S.-E.; Lee, B.; Jeong, B.-R.; Chang, Y.K. Enhancement of biomass and lipid productivity by overexpression of a bZIP transcription factor in *Nannochloropsis salina*. *Biotechnol. Bioeng.* **2017**, *115*, 331–340. [CrossRef]

99. Liang, M.-H.; Jiang, J.-G. Analysis of carotenogenic genes promoters and WRKY transcription factors in response to salt stress in *Dunaliella bardawil*. *Sci. Rep.* **2017**, *7*, 37025. [CrossRef]

100. Matthijs, M.; Fabris, M.; Broos, S.; Vyverman, W.; Goossens, A. Profiling of the Early Nitrogen Stress Response in the Diatom *Phaeodactylum tricornutum* Reveals a Novel Family of RING-Domain Transcription Factors. *Plant Physiol.* **2015**, *170*, 489–498. [CrossRef] [PubMed]

101. Gao, Z.; Li, Y.; Wu, G.; Li, G.; Sun, H.; Deng, S.; Shen, Y.; Chen, G.; Zhang, R.; Meng, C.; et al. Transcriptome Analysis in *Haematococcus pluvialis*: Astaxanthin Induction by Salicylic Acid (SA) and Jasmonic Acid (JA). *PLoS ONE* **2015**, *10*, e0140609. [CrossRef]

102. Tevatia, R.; Allen, J.; Blum, P.; Demirel, Y.; Black, P. Modeling of rhythmic behavior in neutral lipid production due to continuous supply of limited nitrogen: Mutual growth and lipid accumulation in microalgae. *Bioresour. Technol.* **2014**, *170*, 152–159. [CrossRef]

103. Kumar, V.; Muthuraj, M.; Palabhanvi, B.; Das, D. Synchronized growth and neutral lipid accumulation in *Chlorella sorokiniana* FC6 IITG under continuous mode of operation. *Bioresour. Technol.* **2016**, *200*, 770–779. [CrossRef]

104. Wen, X.; Geng, Y.; Li, Y. Enhanced lipid production in *Chlorella pyrenoidosa* by continuous culture. *Bioresour. Technol.* **2014**, *161*, 297–303. [CrossRef] [PubMed]

105. Del Río, E.; García-Gómez, E.; Moreno, J.; Guerrero, M.G.; Garcia-González, M.; Sanchez, E.D.R. Microalgae for oil. Assessment of fatty acid productivity in continuous culture by two high-yield strains, *Chlorococcum oleofaciens* and *Pseudokirchneriella subcapitata*. *Algal Res.* **2017**, *23*, 37–42. [CrossRef]

106. Sobczuk, T.M.; Chisti, Y. Potential fuel oils from the microalga Choricystis minor. *J. Chem. Technol. Biotechnol.* **2010**, *85*, 100–108. [CrossRef]

107. Tang, H.; Chen, M.; Ng, K.S.; Salley, S.O. Continuous microalgae cultivation in a photobioreactor. *Biotechnol. Bioeng.* **2012**, *109*, 2468–2474. [CrossRef] [PubMed]

108. Klok, A.J.; Martens, D.E.; Wijffels, R.H.; Lamers, P.P. Simultaneous growth and neutral lipid accumulation in microalgae. *Bioresour. Technol.* **2013**, *134*, 233–243. [CrossRef] [PubMed]

109. Smith, S.R.; Glé, C.; Abbriano, R.M.; Traller, J.C.; Davis, A.; Trentacoste, E.; Vernet, M.; Allen, A.E.; Hildebrand, M. Transcript level coordination of carbon pathways during silicon starvation-induced lipid accumulation in the diatom *Thalassiosira pseudonana*. *New Phytol.* **2016**, *210*, 890–904. [CrossRef] [PubMed]

110. Zaparoli, M.; Ziemniczak, F.G.; Mantovani, L.; Costa, J.A.V.; Colla, L.M. Cellular Stress Conditions as a Strategy to Increase Carbohydrate Productivity in *Spirulina platensis*. *Bioenergy Res.* **2020**, 1–14. [CrossRef]

111. Jiang, X.-M.; Han, Q.; Gao, X.; Gao, G. Conditions optimising on the yield of biomass, total lipid, and valuable fatty acids in two strains of *Skeletonema menzelii*. *Food Chem.* **2016**, *194*, 723–732. [CrossRef]

112. Brennan, L.; Owende, P. Biofuels from microalgae—A review of technologies for production, processing, and extractions of biofuels and co-products. *Renew. Sustain. Energy Rev.* **2010**, *14*, 557–577. [CrossRef]

113. Rezania, S.; Oryani, B.; Cho, J.; Sabbagh, F.; Rupani, P.F.; Talaiekhozani, A.; Rahimi, N.; Ghahroud, M.L. Technical Aspects of Biofuel Production from Different Sources in Malaysia—A Review. *Processes* **2020**, *8*, 993. [CrossRef]

114. Hu, X.; Liu, B.; Deng, Y.; Bao, X.; Yang, A.; Zhou, J. A novel two-stage culture strategy used to cultivate Chlorella vulgaris for increasing the lipid productivity. *Sep. Purif. Technol.* **2019**, *211*, 816–822. [CrossRef]

115. Rezania, S.; Mohamad, S.E. Response Surface Methodology for Optimization of Ethanol Production from Cocoa Waste. *J. Energy Environ. Pollut.* **2020**, *1*, 7–12.

116. Cheng, D.; Li, D.; Yuan, Y.; Zhou, L.; Li, X.; Wu, T.; Wang, L.; Zhao, Q.; Wei, W.; Sun, Y. Improving carbohydrate and starch accumulation in *Chlorella* sp. AE10 by a novel two-stage process with cell dilution. *Biotechnol. Biofuels* **2017**, *10*, 1–14. [CrossRef]

117. Griffiths, M.J.; Harrison, S.T.L. Lipid productivity as a key characteristic for choosing algal species for biodiesel production. *Environ. Biol. Fishes* **2009**, *21*, 493–507. [CrossRef]

118. Gao, G.; Wu, M.; Fu, Q.; Li, X.; Xu, J. A two-stage model with nitrogen and silicon limitation enhances lipid productivity and biodiesel features of the marine bloom-forming diatom *Skeletonema costatum*. *Bioresour. Technol.* **2019**, *289*, 121717. [CrossRef]

119. Álvarez-Díaz, P.D.; Ruiz, J.; Arbib, Z.; Barragán, J.; Garrido-Pérez, C.; Perales, J.A. Lipid Production of Microalga *Ankistrodesmus falcatus* Increased by Nutrient and Light Starvation in a Two-Stage Cultivation Process. *Appl. Biochem. Biotechnol.* **2014**, *174*, 1471–1483. [CrossRef] [PubMed]

120. Anne-Marie, K.; Yee, W.; Loh, S.H.; Aziz, A.; Cha, T.S. Effects of Excess and Limited Phosphate on Biomass, Lipid and Fatty Acid Contents and the Expression of Four Fatty Acid Desaturase Genes in the Tropical Selenastraceaen *Messastrum gracile* SE-MC4. *Appl. Biochem. Biotechnol.* **2019**, *190*, 1438–1456. [CrossRef] [PubMed]

121. Shokravi, H.; Shaghaghi, T.M. Experimental Study on After-Mixing Temperature Control of Concrete in Batching Plant Production. *IOP Mater. Sci. Eng.* **2020**, *713*. [CrossRef]

122. Ahmad, A.; Osman, S.M.; Cha, T.C.; Loh, S.H. Phosphate-induced changes in fatty acid biosynthesis in *Chlorella* sp. KS-MA2 Strain. *Bio Technologia* **2016**, *97*, 295–304. [CrossRef]

123. Xia, L.; Rong, J.; Yang, H.; He, Q.; Zhang, D.; Hu, C. NaCl as an effective inducer for lipid accumulation in freshwater microalgae *Desmodesmus abundans*. *Bioresour. Technol.* **2014**, *161*, 402–409. [CrossRef]

124. Goh, B.H.H.; Ong, H.C.; Cheah, M.Y.; Chen, W.-H.; Yu, K.L.; Mahlia, T.M.I. Sustainability of direct biodiesel synthesis from microalgae biomass: A critical review. *Renew. Sustain. Energy Rev.* **2019**, *107*, 59–74. [CrossRef]

125. Park, H.; Jung, D.; Lee, J.; Kim, P.; Cho, Y.; Jung, I.; Kim, Z.-H.; Lim, S.-M.; Lee, C.-G. Improvement of biomass and fatty acid productivity in ocean cultivation of *Tetraselmis* sp. using hypersaline medium. *Environ. Biol. Fishes* **2018**, *30*, 2725–2735. [CrossRef]

126. Ra, C.-H.; Kang, C.-H.; Jung, J.-H.; Jeong, G.-T.; Kim, S.-K. Effects of light-emitting diodes (LEDs) on the accumulation of lipid content using a two-phase culture process with three microalgae. *Bioresour. Technol.* **2016**, *212*, 254–261. [CrossRef]

127. Yang, H.; He, Q.; Rong, J.; Xia, L.; Hu, C. Rapid neutral lipid accumulation of the alkali-resistant oleaginous *Monoraphidium dybowskii* LB50 by NaCl induction. *Bioresour. Technol.* **2014**, *172*, 131–137. [CrossRef] [PubMed]

128. Pancha, I.; Chokshi, K.; Maurya, R.; Trivedi, K.; Patidar, S.K.; Ghosh, A.; Mishra, S. Salinity induced oxidative stress enhanced biofuel production potential of microalgae *Scenedesmus* sp. CCNM 1077. *Bioresour. Technol.* **2015**, *189*, 341–348. [CrossRef]

129. Wensel, P.; Helms, G.; Hiscox, B.; Davis, W.C.; Kirchhoff, H.; Bule, M.; Yu, L.; Chen, S. Isolation, characterization, and validation of oleaginous, multi-trophic, and haloalkaline-tolerant microalgae for two-stage cultivation. *Algal Res.* **2014**, *4*, 2–11. [CrossRef]

130. Srinivasan, R.; Mageswari, A.; Subramanian, P.; Suganthi, C.; Chaitanyakumar, A.; Aswini, V.; Gothandam, K.M. Bicarbonate supplementation enhances growth and biochemical composition of *Dunaliella salina* V-101 by reducing oxidative stress induced during macronutrient deficit conditions. *Sci. Rep.* **2018**, *8*, 6972. [CrossRef]

131. Zhao, Y.; Wang, H.-P.; Han, B.; Yu, X. Coupling of abiotic stresses and phytohormones for the production of lipids and high-value by-products by microalgae: A review. *Bioresour. Technol.* **2019**, *274*, 549–556. [CrossRef] [PubMed]

132. Guldhe, A.; Renuka, N.; Singh, P.; Bux, F. Effect of phytohormones from different classes on gene expression of *Chlorella sorokiniana* under nitrogen limitation for enhanced biomass and lipid production. *Algal Res.* **2019**, *40*, 101518. [CrossRef]

133. Sivaramakrishnan, R.; Incharoensakdi, A. Enhancement of total lipid yield by nitrogen, carbon, and iron supplementation in isolated microalgae. *J. Phycol.* **2017**, *53*, 855–868. [CrossRef] [PubMed]

134. Zhao, Y.; Li, D.; Xu, J.-W.; Zhao, P.; Li, T.; Ma, H.; Yu, X. Melatonin enhances lipid production in *Monoraphidium* sp. QLY-1 under nitrogen deficiency conditions via a multi-level mechanism. *Bioresour. Technol.* **2018**, *259*, 46–53. [CrossRef]

135. Che, R.; Ding, K.; Huang, L.; Zhao, P.; Xu, J.-W.; Li, T.; Ma, H.; Yu, X. Enhancing biomass and oil accumulation of *Monoraphidium* sp. FXY-10 by combined fulvic acid and two-step cultivation. *J. Taiwan Inst. Chem. Eng.* **2016**, *67*, 161–165. [CrossRef]

136. Babu, A.G.; Wu, X.; Kabra, A.N.; Kim, D.-P. Cultivation of an indigenous *Chlorella sorokiniana* with phytohormones for biomass and lipid production under N-limitation. *Algal Res.* **2017**, *23*, 178–185. [CrossRef]

137. Su, C.-H.; Chien, L.-J.; Gomes, J.; Lin, Y.-S.; Yu, Y.-K.; Liou, J.-S.; Syu, R.-J. Factors affecting lipid accumulation by *Nannochloropsis oculata* in a two-stage cultivation process. *Environ. Biol. Fishes* **2010**, *23*, 903–908. [CrossRef]

138. Abomohra, A.E.-F.; El-Naggar, A.H.; Alaswad, S.O.; Elsayed, M.; Li, M.; Li, W. Enhancement of biodiesel yield from a halophilic green microalga isolated under extreme hypersaline conditions through stepwise salinity adaptation strategy. *Bioresour. Technol.* **2020**, *310*, 123462. [CrossRef]

139. Hang, L.T.; Mori, K.; Tanaka, Y.; Morikawa, M.; Toyama, T. Enhanced lipid productivity of *Chlamydomonas reinhardtii* with combination of NaCl and $CaCl_2$ stresses. *Bioprocess Biosyst. Eng.* **2020**, *43*, 971–980. [CrossRef] [PubMed]

140. Mirizadeh, S.; Nosrati, M.; Shojaosadati, S.A. Synergistic Effect of Nutrient and Salt Stress on Lipid Productivity of *Chlorella vulgaris* through Two-Stage Cultivation. *BioEnergy Res.* **2019**, *13*, 507–517. [CrossRef]

141. Acién, F.; Molina, E.; Reis, A.; Torzillo, G.; Zittelli, G.; Sepúlveda, C.; Masojídek, J. Photobioreactors for the production of microalgae. In *Microalgae-Based Biofuels and Bioproducts*; Elsevier: Cambridge, UK, 2017; pp. 1–44.

142. Yen, H.-W.; Chang, J.-T. A two-stage cultivation process for the growth enhancement of *Chlorella vulgaris*. *Bioprocess Biosyst. Eng.* **2013**, *36*, 1797–1801. [CrossRef]

143. Cheah, W.Y.; Show, P.-L.; Juan, J.C.; Chang, J.-S.; Ling, T.C. Enhancing biomass and lipid productions of microalgae in palm oil mill effluent using carbon and nutrient supplementation. *Energy Convers. Manag.* **2018**, *164*, 188–197. [CrossRef]

144. Caprio, F.D.; Andrea, V.; Altimari, P.; Toro, L.; Masciocchi, B.; Iaquaniello, G.; Pagnanelli, F. Two stage process of microalgae cultivation for starch and carotenoid production. *Chem. Eng. Trans.* **2016**, *49*, 415–420.

145. Ra, C.H.; Kang, C.-H.; Kim, N.K.; Lee, C.-G.; Kim, S.-K. Cultivation of four microalgae for biomass and oil production using a two-stage culture strategy with salt stress. *Renew. Energy* **2015**, *80*, 117–122. [CrossRef]

146. Chen, G.; Zhao, L.; Qi, Y. Enhancing the productivity of microalgae cultivated in wastewater toward biofuel production: A critical review. *Appl. Energy* **2015**, *137*, 282–291. [CrossRef]

147. Bora, A.P.; Gupta, D.P.; Durbha, K.S. Sewage sludge to bio-fuel: A review on the sustainable approach of transforming sewage waste to alternative fuel. *Fuel* **2020**, *259*, 116262. [CrossRef]

148. Xiong, W.; Gao, C.; Yan, N.; Wu, C.; Wu, Q. Double CO_2 fixation in photosynthesis–fermentation model enhances algal lipid synthesis for biodiesel production. *Bioresour. Technol.* **2010**, *101*, 2287–2293. [CrossRef]

149. Zhou, W.; Hu, B.; Li, Y.; Min, M.; Mohr, M.; Du, Z.; Chen, P.; Ruan, R. Mass Cultivation of Microalgae on Animal Wastewater: A Sequential Two-Stage Cultivation Process for Energy Crop and Omega-3-Rich Animal Feed Production. *Appl. Biochem. Biotechnol.* **2012**, *168*, 348–363. [CrossRef] [PubMed]

150. Zhou, W.; Min, M.; Li, Y.; Hu, B.; Ma, X.; Cheng, Y.; Liu, Y.; Chen, P.; Ruan, R. A hetero-photoautotrophic two-stage cultivation process to improve wastewater nutrient removal and enhance algal lipid accumulation. *Bioresour. Technol.* **2012**, *110*, 448–455. [CrossRef] [PubMed]

151. Liu, X.; Hong, Y.; Liu, P.; Zhan, J.; Yan, R. Effects of cultivation strategies on the cultivation of *Chlorella* sp. HQ in photoreactors. *Front. Environ. Sci. Eng.* **2019**, *13*, 78. [CrossRef]

152. Jagadevan, S.; Banerjee, C.; Banerjee, C.; Guria, C.; Tiwari, R.; Baweja, M.; Shukla, P. Recent developments in synthetic biology and metabolic engineering in microalgae towards biofuel production. *Biotechnol. Biofuels* **2018**, *11*, 1–21. [CrossRef]

153. Wang, X.-W.; Huang, L.; Ji, P.-Y.; Chen, C.-P.; Li, X.-S.; Gao, Y.-H.; Liang, J.-R. Using a mixture of wastewater and seawater as the growth medium for wastewater treatment and lipid production by the marine diatom *Phaeodactylum tricornutum*. *Bioresour. Technol.* **2019**, *289*, 121681. [CrossRef]

154. Farooq, W.; Lee, Y.-C.; Ryu, B.-G.; Kim, B.-H.; Kim, H.-S.; Choi, Y.-E.; Yang, J.-W. Two-stage cultivation of two *Chlorella* sp. strains by simultaneous treatment of brewery wastewater and maximizing lipid productivity. *Bioresour. Technol.* **2013**, *132*, 230–238. [CrossRef] [PubMed]

155. Hena, S.; Fatihah, N.; Tabassum, S.; Ismail, N. Three stage cultivation process of facultative strain of *Chlorella sorokiniana* for treating dairy farm effluent and lipid enhancement. *Water Res.* **2015**, *80*, 346–356. [CrossRef] [PubMed]

156. Ma, X.; Zheng, H.; Addy, M.; Anderson, E.; Liu, Y.; Chen, P.; Ruan, R. Cultivation of Chlorella vulgaris in wastewater with waste glycerol: Strategies for improving nutrients removal and enhancing lipid production. *Bioresour. Technol.* **2016**, *207*, 252–261. [CrossRef]

157. Ge, S.; Qiu, S.; Tremblay, D.; Viner, K.; Champagne, P.; Jessop, P.G. Centrate wastewater treatment with *Chlorella vulgaris*: Simultaneous enhancement of nutrient removal, biomass and lipid production. *Chem. Eng. J.* **2018**, *342*, 310–320. [CrossRef]

158. Johnson, T.J.; Katuwal, S.; Anderson, G.A.; Gu, L.; Zhou, R.; Gibbons, W.R. Photobioreactor cultivation strategies for microalgae and cyanobacteria. *Biotechnol. Prog.* **2018**, *34*, 811–827. [CrossRef] [PubMed]

159. Sirisuk, P.; Ra, C.-H.; Jeong, G.-T.; Kim, S.-K. Effects of wavelength mixing ratio and photoperiod on microalgal biomass and lipid production in a two-phase culture system using LED illumination. *Bioresour. Technol.* **2018**, *253*, 175–181. [CrossRef]

160. Ramanna, L.; Rawat, I.; Bux, F. Light enhancement strategies improve microalgal biomass productivity. *Renew. Sustain. Energy Rev.* **2017**, *80*, 765–773. [CrossRef]

161. Ra, C.H.; Kang, C.-H.; Jung, J.-H.; Jeong, G.-T.; Kim, S.-K. Enhanced biomass production and lipid accumulation of *Picochlorum atomus* using light-emitting diodes (LEDs). *Bioresour. Technol.* **2016**, *218*, 1279–1283. [CrossRef]

162. Huerlimann, R.; Denys, R.; Heimann, K. Growth, lipid content, productivity, and fatty acid composition of tropical microalgae for scale-up production. *Biotechnol. Bioeng.* **2010**, *107*, 245–257. [CrossRef]

163. Wang, T.; Tian, X.; Liu, T.; Wang, Z.-J.; Guan, W.; Guo, M.; Chu, J.; Zhuang, Y. A two-stage fed-batch heterotrophic culture of *Chlorella protothecoides* that combined nitrogen depletion with hyperosmotic stress strategy enhanced lipid yield and productivity. *Process. Biochem.* **2017**, *60*, 74–83. [CrossRef]

164. Yoon, S.Y.; Hong, M.E.; Chang, W.S.; Sim, S.J. Enhanced biodiesel production in *Neochloris oleoabundans* by a semi-continuous process in two stage photobioreactors. *Bioprocess Biosyst. Eng.* **2015**, *38*, 1415–1421. [CrossRef]

165. Nayak, M.; Suh, W.I.; Cho, J.M.; Kim, H.S.; Lee, B.; Chang, Y.K. Strategic implementation of phosphorus repletion strategy in continuous two-stage cultivation of *Chlorella* sp. HS2: Evaluation for biofuel applications. *J. Environ. Manag.* **2020**, *271*, 111041. [CrossRef]
166. Ghidossi, T.; Marison, I.; Devery, R.; Gaffney, D.; Forde, C. Characterization and Optimization of a Fermentation Process for the Production of High Cell Densities and Lipids Using Heterotrophic Cultivation of *Chlorella protothecoides*. *Ind. Biotechnol.* **2017**, *13*, 253–259. [CrossRef]
167. Palabhanvi, B.; Muthuraj, M.; Mukherjee, M.; Kumar, V.; Das, D. Process engineering strategy for high cell density-lipid rich cultivation of *Chlorella* sp. FC2 IITG via model guided feeding recipe and substrate driven pH control. *Algal Res.* **2016**, *16*, 317–329. [CrossRef]
168. Zheng, Y.; Li, T.; Yu, X.; Bates, P.D.; Dong, T.; Chen, S. High-density fed-batch culture of a thermotolerant microalga *Chlorella sorokiniana* for biofuel production. *Appl. Energy* **2013**, *108*, 281–287. [CrossRef]
169. Muylaert, K.; Bastiaens, L.; Vandamme, D.; Gouveia, L. Harvesting of microalgae: Overview of process options and their strengths and drawbacks. In *Microalgae-Based Biofuels and Bioproducts*; Elsevier: Cambridge, UK, 2017; pp. 113–132.
170. Zhang, B.; Chen, S. Effect of different organic matters on flocculation of *Chlorella sorokiniana* and optimization of flocculation conditions in swine manure wastewater. *Bioresour. Technol.* **2015**, *192*, 774–780. [CrossRef] [PubMed]
171. Chatsungnoen, T.; Chisti, Y. Oil production by six microalgae: Impact of flocculants and drying on oil recovery from the biomass. *Environ. Biol. Fishes* **2016**, *28*, 2697–2705. [CrossRef]
172. Vandamme, D.; Foubert, I.; Fraeye, I.; Muylaert, K. Influence of organic matter generated by Chlorella vulgaris on five different modes of flocculation. *Bioresour. Technol.* **2012**, *124*, 508–511. [CrossRef] [PubMed]
173. Pandey, A.; Shah, R.; Yadav, P.; Verma, R.; Srivastava, S. Harvesting of freshwater microalgae *Scenedesmus* sp. by electro–coagulation–flocculation for biofuel production: Effects on spent medium recycling and lipid extraction. *Environ. Sci. Pollut. Res.* **2019**, *27*, 3497–3507. [CrossRef]
174. Fayad, N.; Yehya, T.; Audonnet, F.; Vial, C. Harvesting of microalgae *Chlorella vulgaris* using electro-coagulation-flocculation in the batch mode. *Algal Res.* **2017**, *25*, 1–11. [CrossRef]
175. Mathimani, T.; Mallick, N. A comprehensive review on harvesting of microalgae for biodiesel—Key challenges and future directions. *Renew. Sustain. Energy Rev.* **2018**, *91*, 1103–1120. [CrossRef]
176. Vashist, V.; Chauhan, D.; Bhattacharya, A.; Rai, M.P. Role of silica coated magnetic nanoparticle on cell flocculation, lipid extraction and linoleic acid production from *Chlorella pyrenoidosa*. *Nat. Prod. Res.* **2019**, *34*, 2852–2856. [CrossRef]
177. Ahmad, A.L.; Yasin, N.M.; Derek, C.; Lim, J. Comparison of harvesting methods for microalgae *Chlorella* sp. and its potential use as a biodiesel feedstock. *Environ. Technol.* **2014**, *35*, 2244–2253. [CrossRef]
178. Borges, L.; Caldas, S.; D'Oca, M.G.M.; Abreu, P.C. Effect of harvesting processes on the lipid yield and fatty acid profile of the marine microalga *Nannochloropsis oculata*. *Aquac. Rep.* **2016**, *4*, 164–168. [CrossRef]
179. Yasin, N.H.M.; Shafei, N.I.; Rushan, N.H.; Abu Sepian, N.R.; Said, F.M. The Effect of Microalgae Harvesting on Lipid for Biodiesel Production. *Mater. Today Proc.* **2019**, *19*, 1582–1590. [CrossRef]
180. Coward, T.; Lee, J.; Caldwell, G.S. Harvesting microalgae by CTAB-aided foam flotation increases lipid recovery and improves fatty acid methyl ester characteristics. *Biomass Bioenergy* **2014**, *67*, 354–362. [CrossRef]
181. Magnusson, M.; Mata, L.; Denys, R.; Paul, N. Biomass, Lipid and Fatty Acid Production in Large-Scale Cultures of the Marine Macroalga *Derbesia tenuissima* (Chlorophyta). *Mar. Biotechnol.* **2014**, *16*, 456–464. [CrossRef]
182. Sekar, M.; Mathimani, T.; Alagumalai, A.; Chi, N.T.L.; Duc, P.A.; Bhatia, S.K.; Brindhadevi, K.; Pugazhendhi, A. A review on the pyrolysis of algal biomass for biochar and bio-oil-Bottlenecks and scope. *Fuel* **2021**, *283*, 119190. [CrossRef]

Publisher's Note: MDPI stays neutral with regard to jurisdictional claims in published maps and institutional affiliations.

Article

An Extensive Analysis of Biodiesel Blend Combustion Characteristics under a Wide-Range of Thermal Conditions of a Cooperative Fuel Research Engine

Vu H. Nguyen [1], Minh Q. Duong [2], Kien T. Nguyen [1], Thin V. Pham [3] and Phuong X. Pham [1,*]

[1] Department of IC Engine, Faculty of Vehicle & Energy Engineering, Le Quy Don Technical University, 236 Hoang Quoc Viet, Bac Tu Liem, Hanoi 10000, Vietnam; vunguyenhoang@lqdtu.edu.vn (V.H.N.); kiennt@lqdtu.edu.vn (K.T.N.)

[2] Division of Automobile Engineering Technology, Faculty of Mechanical Engineering, University of Transport Technology, Hanoi 10000, Vietnam; minhdq@utt.edu.vn

[3] Department of Physics, Faculty of Physical & Chemical Engineering, Le Quy Don Technical University, 236 Hoang Quoc Viet, Bac Tu Liem, Hanoi 10000, Vietnam; thinpv@lqdtu.edu.vn

* Correspondence: phuongpham@lqdtu.edu.vn

Received: 22 June 2020; Accepted: 30 July 2020; Published: 17 September 2020

Abstract: Examining the influence of thermal conditions in the engine cylinder on engine combustion characteristics is critically important. This may help to understand physical and chemical processes occurring in engine cycles and this is relevant to both fossil fuels and alternative fuels like biodiesels. In this study, six different biodiesel–diesel blends (B0, B10, B20, B40, B60 and B100 representing 0, 10, 20, 40, 60 and 100% by volume of biodiesel in the diesel–biodiesel mixtures, respectively) have been successfully tested in a cooperative fuel research (CFR) engine operating under a wide range of thermal conditions at the start of fuel injection. This is a standard cetane testing CFR-F5 engine, a special tool for fuel research. In this study, it was further retrofitted to investigate combustion characteristics along with standard cetane measurements for those biodiesel blends. The novel biodiesel has been produced from residues taken from a palm cooking oil manufacturing process. It is found that the cetane number of B100 is almost 30% higher than that of B0 and this could be attributed to the oxygen content in the biofuel. Under similar thermal conditions at the start of injection, it is observed that the influence of engine load on premixed combustion is minimal. This could be attributable to the well-controlled intake air temperature in this special engine and therefore the evaporation and mixing rate prior to the start of combustion is similar under different loading conditions. Owing to higher cetane number (CN), B100 is more reactive and auto-ignites up to 3 degrees of crank angle (DCA) earlier compared to B0. It is generally observed in this study that B10 shows a higher maximum value of in-cylinder pressure compared to that of B0 and B20. This could be evidence for lubricant enhancement when operating the engine with low-blending ratio mixtures like B10 in this case.

Keywords: biodiesel; CFR engine; engine combustion; ignition delay; maximum rate of pressure rise; thermal condition at start of injection

---

1. Introduction

Biodiesels are environmentally friendly alternative fuels and have important properties close to that of fossil diesel. Biodiesels are manufactured from different feedstock including vegetable oils, animal fats and algae [1]. The transesterification reaction between triglycerides available in the feedstock and methanol is the standard method to produce biodiesels (a mixture of different fatty acid methyl esters—FAMEs) and glycerols [1,2]. A number of studies on utilizations of biodiesels and their blends in compression ignition engines can be found in current literature [2–8]. Excellent reviews on the topic (both fundamental studies and biodiesels' utilizations) are provided in [9–12]. Key findings are: (i) Biodiesels have lower calorific values compared to fossil counterparts and this leads to a penalty in fuel economy when operating engines with biodiesels and their blends. The increase in fuel consumption when using biodiesels is approximately proportional to the loss in their calorific value [1,13,14]; (ii) compared to diesel, blends of biodiesel–diesel normally show shorter ignition delay times and a reduced heat release rate (HRR) as well as a slightly higher efficiency [10]. The shorter ignition delay times are due to the higher cetane number (CN) of biodiesels; (iii) significant reductions in soot emission compositions in the engine exhaust, while there are contradictory statements regarding particle size distributions and NOx emissions [9]. It is suggested by Damanik et al. in [11] that the trends in engine performance and emission levels when operating with biodiesel blends should be interpreted with caution as generalization of the trends is not possible using the results currently available in the literature [11]. Apart from controversies in some of the reported results as mentioned above, there are few key issues with respect to existing research on biodiesels: (i) the fuels are generally selected at random depending on their availability and without any reference to chemical and physical property variations amongst biodiesels derived from different feedstock; (ii) very few studies report on the basic auto-ignitability of biodiesels, a feature that may well be important considering how the influence of different thermal conditions in the cylinder at the start of injection (SOI) on the in-cylinder pressure development, heat release rate and ignition delay provides a better understanding of biodiesel auto-ignitability; and (iii) the definition of engine load condition is not consistent amongst the literature, a number of authors used similar fuel volume flow rates (mL/min) while some others used a similar amount of input energy (MJ/min) when testing auto-ignition engines with biodiesel blends and diesel. Those confusions may make the comparison of engine performance when operating with diesel and biodiesel blends irrelevant.

The engine cycle is very complex as it includes a number of physical processes (e.g., atomization, evaporation and mixing) as well as chemical processes (e.g., auto-ignition and combustion). Therefore, studies on these processes are normally conducted in many different laboratory tools such as open burners [15,16], shock tubes [17], single cylinder engines [18–22] and multi-cylinder engines [14,23]. Fundamental tools like pilot and co-flow burners [24,25] can be used to deeply investigate an isolated process like primary atomization [25,26], secondary atomization [27] or auto-ignition [28]. Single-cylinder engines add more complex processes to that occurring in fundamental tools (e.g., laboratory burners, shock tubes and rapid compression machine), and single engines like the CFR engine used in this work have been shown as a useful equipment to closely describe the engine cycle. Single-cylinder engines normally have a capability of varying compression ratio (CR) that is impossible in practical multi-cylinder engines [20,29]. In an engine cycle, investigating thermal conditions at SOI is critically important as the conditions strongly affect the ignition delay, fuel–air premixed fraction, in-cylinder pressure development and, as such, engine power and efficiency. Studies conducted in burners, shock tubes, single-cylinder engines and multi-cylinder engines need to address this.

Cooperative fuel research (CFR) engines have been developed for fuel testing. The special tools are single-cylinder and variable CR engines. The engines were initially used for examining fuels but are now used worldwide for exploring the combustion characteristics of research fuels under one of the five methods: the motor, research, aviation, supercharge and cetane methods [30]. Model F-1/F-2 Combination CFR engine is used to determine the fuel octane number of gasoline-like fuels. This testing method is conducted under ASTM D2699 and D2700 standards. Model CFR F-5 engine (the one used in this study) is a complete system for measuring the CN of diesel-like fuels, conforming to the ASTM D613 standard. This method is accepted worldwide as the standard for determining the auto-ignition quality of diesel fuels. Further details of the CFR engine used in this study will be shown later in Section 2.1. Using these special engines to investigate combustion characteristics of biodiesel and its blends is relevant. The capability of varying the compression ratio makes them special for fuel testing.

Using a variable CR single-cylinder engine, blends of castor oil-based biodiesel and fossil diesel (B0, B10, B20, B30, B40, B50 and B100) have been tested in [31] to investigate the influence of CR and blending ratio on mean gas temperature, cylinder pressure variation, net heat release and mass fraction burned. The authors have observed that an increase in CR leads to an increase in mean gas temperature and a decrease in net heat release rate. Another study done by Dash et al. [32] in a variable CR single-cylinder engine operating with different Nahar biodiesel blends (B0, B5, B10, B20, B30, B40 and B50) showed that combustion duration reduces when increasing the blending ratio up to B40, then slightly increases for B50 and significantly for B100. It was also observed that the blending ratio has significant effects on the maximum rate of pressure rise (MRPR), heat release rate and ignition delay. The influence of thermal conditions at SOI is not addressed in these studies [31,32].

The performance of a four-cylinder engine operating with waste oil, rapeseed oil and corn oil biodiesels and diesel has been investigated by Tesfa et al. [33] and showed that the influence of fuel types on heat release rate and specific fuel consumption is not significant. Pham et al. have tested a number of biodiesels having different physic-chemical profiles in a single-cylinder engine [7,21], a multiple-cylinder engine [5,34] and burners [24,25], and the authors claimed that the molecular profiles of biodiesels, determined by their feedstock, have a significant impact on atomization and combustion characteristics.

Although a number of modern techniques including laser diagnosis can be utilized to study engine combustion as noted briefly above, in-cylinder pressure transducers have been shown as one of the most convenient and efficient tools to investigate the engine cycle including combustion characteristics. Certainly, lasers are a very powerful technique that can be used to deeply diagnose the physical chemical processes such as quantifying auto-ignition zone [35], measuring combustion radicals (OH-, and formaldehyde), flame structure and emission. The measurements are impossible using pressure transducers. Pressure transducers, however, are much cheaper and much easier to setup and operate. Pressure signals have been used to investigate net heat release, thermal efficiency, air mass and fuel flow rates, in-cylinder trapped mass, exhaust gas recirculation, emission and noise control [36–40].

It was reported earlier by Vargas et al. [9] that the number of studies on biodiesels have been increased rapidly in the last decade. Extensive investigations of in-cylinder pressure development under a wide range of cylinder thermal conditions at SOI, according to the authors' knowledge, are scarce in the literature.

In this study, a CFR F5 engine, a standardized machine known as a cetane testing engine, was firstly employed with ASTM-D613 [41] to measure CN for different biodiesel–diesel blends (B0, B10, B20, B40, B60 and B100). The biodiesel used here is produced from the residue of a cooking oil manufacturing process (not used cooking oil). Then, the system was further equipped with a fast-response in-cylinder pressure transducer and an encoder to measure the in-cylinder pressure of the engine. This extension aims to provide an extensive examination of the in-cylinder pressure development under a wide range of thermal conditions in the cylinder at the timing of fuel injection. The injection timing and CR are varied in this study so that the thermal conditions at SOI are varied in a wide range.

2. Experiment Setup and Testing Conditions

2.1. Experiment Description

The experiment system used in this study is schematically described in Figure 1a (for the cooperative fuel research—CFR engine test bed) and Figure 1b (for the enlarged combustion chamber of the CFR engine). This is a CFR-F5 engine designed for cetane testing. As mentioned earlier, CFR engines are capable of varying CR and this makes them special, especially for testing combustion characteristics under different thermal conditions like the one conducted in this study. Under the standardized operating condition of the engine to measure liquid fuel CN [41], the CFR engine is operated with a speed of 900 rpm, a CR of 13 and a well-controlled air intake temperature of 65 °C for all testing conditions. This temperature is quite high compared to that in conventional engines. The fuel is supplied into the cylinder through an injector (#8) under an injection timing of 13 degrees of crank angle (DCA). Ignition delay, used for computing CN, is measured using a combustion pickup sensor (#7) along with a delay sensor (#9).

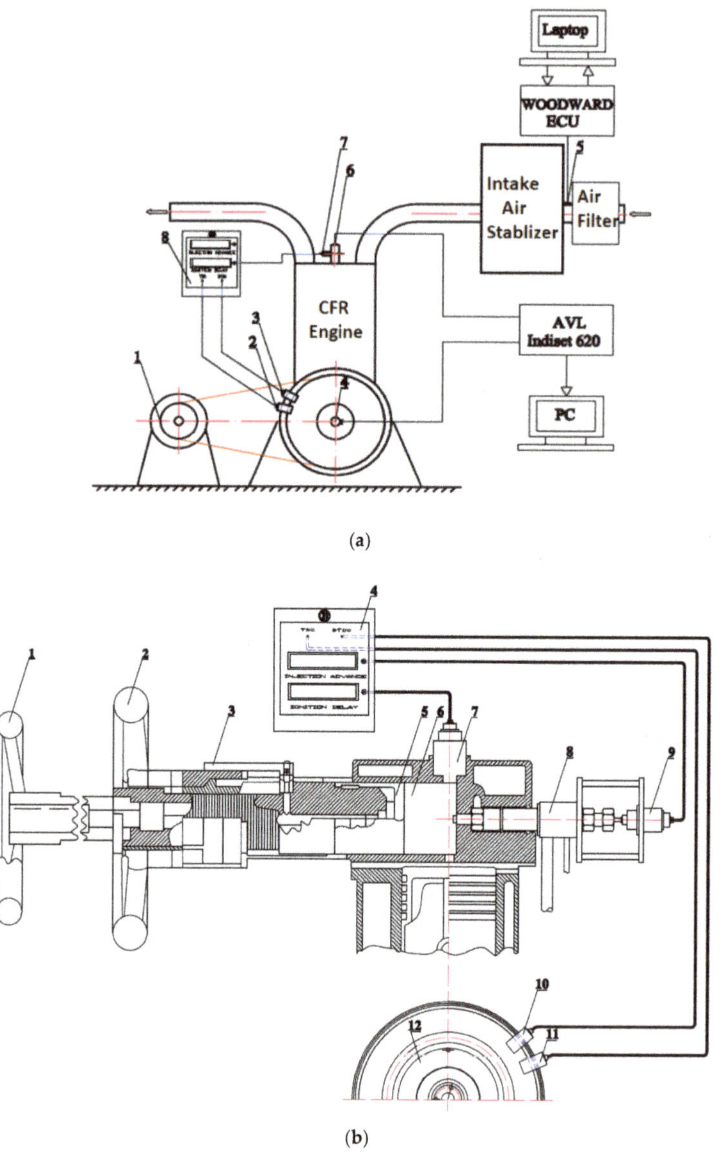

(a)

(b)

Figure 1. Experiment setup. (**a**) Cooperative fuel research (CFR) engine test-bed: 1. electric dyno; 2. top dead center (TDC) pickup sensor; 3. 13 degrees of crank angle (DCA) pickup sensor; 4. encoder; 5. air flow sensor; 6.pressure sensor; 7. combustion pickup sensor; (**b**) the CFR engine's combustion chamber: 1. locking wheel; 2. handwheel; 3. micrometer; 4. ignition delay meter; 5. piston; 6. combustion chamber; 7. combustion pickup sensor; 8. injector; 9. injector needle lift sensor; 10. TDC pickup sensor; 11. 13 DCA pickup sensor; 12. flywheel.

Different from conventional engines, at a specific CR, the combustion chamber of the CFR engine shown in Figure 1 is a constant volume chamber (#6 shown in Figure 1b) located above the piston and separately from the chamber formed by the cylinder, piston and cylinder head (chamber 2). The CR varies by changing the volume of the chamber and this is done by rotating the handwheel (#2 shown in Figure 1b). The position of the piston is determined using the micrometer (#3 shown in Figure 1b). The fuel injection timing is exactly measured using an injector needle lift sensor (#9 shown in Figure 1b). The start of combustion timing is measured using a combustion pickup sensor (#7 shown in Figure 1b). The combustion chamber is connected to chamber 2 by a small hole located in the piston's centerline as shown in Figure 1a. Due to these special characteristics, combustion in this engine mainly occurs in the constant volume chamber, chamber 2 just works like a pumping system to create high-pressure conditions for the combustion chamber. This special equipment is good for studies on combustion characteristics like in-cylinder pressure development, heat release rate and auto-ignitability. As the combustion chamber of the CFR engine is a constant volume chamber and has a high air intake temperature as mentioned, this tool may not be suitable for studies related to whole engine cycles in which emission concentrations, engine power, fuel economy and other engine performance characteristics are examined.

In this study, the engine is further extended to include a fast-response in-cylinder pressure transducer (#6, AVL QC33C), an encoder (#4) and a data acquisition and control system (AVL Indiset 620). The extension also includes a gas flow sensor to measure the amount of intake air and an equivalent air–fuel (or lambda) sensor equipped with a control unit, ECM-0565-128-0702-C manufactured by WOODWARD as shown schematically in Figure 1. CFR engine specifications are presented in Table 1.

Table 1. CFR-F5 engine operating conditions under cetane testing modes [7].

Parameters, [Unit]	Value	Note
Speed, [rpm]	900 ± 9	Rpm—revolution per minute
Injection timing, [DCA BTDC]	13	DCA—degree of crank angle BTDC—before top dead center
Reference ignition delay, [DCA BTDC]	13	
Fuel line injection pressure, [bar]	103.0 ± 0.2	
Compression ratio	8 ÷ 36	
Coolant temperature for injector, [°C]	38 ± 3	
Lubricant pressure, [bar]	1.75 ÷ 2.10	
Lubricant temperature, [°C]	57 ± 8	
Engine coolant temperature, [°C]	100 ± 2	
Intake air temperature, [°C]	65 ± 0.5	
Valve thermal gaps, [mm]	0.200 ± 0.025	
Lubricant type	SAE 30	

2.2. Fuels Tested

The biodiesel tested here is produced from the residue of a palm cooking oil production process and this was reported in our previous study [42]. It was found that the residue left from the cooking oil production process (not used cooking oil) is still rich in fatty acid esters. Then, biodiesel used in this work was successfully derived from the special feedstock using triple cycles of heterogeneous catalyzed transesterification [7]. Fuel blends used here include B0 (pure diesel), B10, B20, B40, B60 and B100 (pure biodiesel) corresponding to 0, 10, 20, 40, 60 and 100 vol% of biodiesel in the biodiesel–diesel blends, respectively. Important properties of diesel and biodiesel fuels are summarized in Table 2. It is noted here that important properties of these blends have been carefully measured and reported

Sustainability **2020**, *12*, 7666

elsewhere in our previous study [42]. Important physicochemical properties of all blends tested here have been carefully measured using relevant testing methods, as shown in Table 2.

Table 2. Important properties of fuels tested [7].

Property	Unit	Testing Method	B100	B60	B40	B20	B10	D(B0)
Ester content	wt%	EN 14103	98.91	—	—	—	—	—
Glycerin content	wt%	ASTM D6584	0.0	—	—	—	—	—
Phosphorus content	wt%	ASTM D4951	0.0002	—	—	—	—	—
Sodium/potassium, combined	mg/kg	EN 14538	0.1	—	—	—	—	—
Oxidation stability, 0 months	h	EN 14112	6.02	—	—	24.07	111.9	—
Oxidation stability, 8 months	h	EN 14112	—	—	—	6.8	87	—
Palmitic, C16:0	wt%	—	28.09	—	—	—	—	—
Stearic, C18:0	wt%	—	9.53	—	—	—	—	—
Oleic, C18:1	wt%	—	43.47	—	—	—	—	—
Linoleic, C18:2	wt%	—	18.02	—	—	—	—	—
Iodine value	gI/100 g	EN 14111	48.0	—	—	—	—	—
Saponification number	mgKOH/g	ASTM D664-04	177.3	—	—	—	—	—
Acid number	mgKOH/g	ASTM D664	0.06	—	—	—	—	—
Water content	wt%	ASTM D95-05	0.20	—	—	—	—	—
Flash point	°C	ASTM D93	183.5	130.3	110.5	87.80	77.08	68.50
Kinematic viscosity at 40 °C	mm^2/s	ASTM D445	4.6	3.85	3.60	3.32	3.18	3.11
Relative density at 15 °C	—	ASTM D1298	0.874	0.865	0.852	0.845	0.842	0.839
Higher heating values	MJ/kg	—	38.10	41.53	42.01	44.62	45.20	46.18
Cloud point	°C	ASTM D2500	18	—	—	—	—	—
Pour point	°C	ASTM D97	—	—	—	−3	−3	—
Cetane number	—	ASTM D613	66.9	62.4	57.4	54.5	53.7	52.4
Auto ignition temperature	K	ASTM E659-78	494	—	—	—	—	481
Molecular weight	g/mol		295.31	243.86	223.91	206.74	199.02	191.8
Atom Fraction								
C content	wt%		76.96	80.86	82.85	84.87	85.90	86.93
H content	wt%		12.17	12.48	12.64	12.80	12.88	12.96
O content	wt%		10.83	6.62	4.47	2.29	1.18	0.07

It is noted that the molecular structure of one biodiesel solely depends on its mother feedstock. Empirical correlations are developed to correlate these relevant properties to the fuel structure using parameters such as iodine values (IV) and saponification number (SN). The IV is the number of grams of iodine consumed per 100 g of fatty acid. It is being used as a measure of unsaturation levels in fatty acid (a higher IV indicates a higher degree of unsaturation). The SN is the mass of potassium hydroxide (KOH) required to saponify 1 g of FAME; therefore, SN reflects the carbon chain length (a higher SN implies a shorter carbon chain length). IV and SN of the palm oil-based biodiesel were carefully measured using standards EN 14,111 and ASTM D664-04, respectively. As shown in Table 2, the biodiesel (B100) tested in this study has a medium IV (IV = 48) and a high saponification number (SN = 148) and this means that the fuel has a long carbon chain length and high unsaturation degree. The C/H/O values reported in Table 2 for the pure biodiesel (B100) and pure diesel (B0) are carefully measured using high-performance liquid chromatography (HPLC). C/H/O values of blends are calculated using the blending ratio and C/H/O values of B100 and B0.

One biodiesel may have constituents with 8 to 25 carbon numbers and up to 5 or even more numbers of double bonds [5] but the molecule always has two oxygen atoms, and this key feature makes the fuels different from the fossil diesel counterpart. The fuel oxygen content (FOC) in biodiesels may enhance the fuel reactivity [43]. This is critically important because of the local rich fuel–air mixture (lack of oxygen) in the auto-ignition zone [35,43] of compression ignition engines. The cetane number of the fuels shown in Table 2 is tested using the CFR engine operating with the standardized approach [41]. It is clearly shown from Table 2 that CN increases from B0 to B100 and this may be attributed to the oxygen content in the blends [43]. The cetane number of pure biodiesel is almost 30% higher than that of fossil diesel. Oxygen content in biodiesel blends may make the fuel–air mixture leaner in the auto-ignition zone and this in turn enhances the fuel reactivity.

2.3. Testing Points

The original facilities in conjunction with the additional items described above (Sections 2.1 and 2.2) extend the capability of the test bed to measure in-cylinder pressure signals under a wide range of air–fuel equivalent ratios (λ), compression ratios and injection timings. In a nutshell, tests conducted in this study include (i) standard CN tests for all biodiesel blends (B0, B10, B20, B40, B40 and B100) and (ii) tests of those blends under different thermal conditions in the cylinder at SOI. Under these injection timing conditions, the injection temperature, Ti, varies between 775 and 865 K under CR = 15. Under CR = 17, Ti ranges from 790 to 890 K. These values of Ti are calculated using in-cylinder pressure at SOI, pi, which is the output from the in-cylinder pressure signals experimentally measured. The temperature conditions cover a wide range of thermal conditions at SOI in CI engines.

The CN tests are conducted (ASTM D613) under an engine speed of 900 rpm; an injection timing of 13 DCA before top dead center (BTDC); an amount of fuel injection of 13 ± 0.02 mL/min; and a well-controlled intake air temperature, Ti, of 65 °C. These conditions and others related to CN testing are reported in Table 1.

The engine conditions used for part (ii) mentioned above can be summarized here:

(1)   Engine speed: 900 rpm, similar to the standardized measurement for CN;

(2)   Intake air temperature: well controlled to remain constant at 65 °C, similar to the standardized measurement of CN;

(3)   Two compression ratios (CR): 15 and 17; different from the standardized measurement for CN;

(4)   Seven injection timings, t_{inj}: 8, 10, 12, 14, 16, 18 and 20 DCA before TDC; different from the standardized measurement for CN;

(5)   Four fuel flow rates: 15.5, 13.0, 11.30 and 10.0 mL/min. These flow rate conditions are described in this work as M1, M2, M3 and M4, respectively. The M1 to M4 conditions are applied to all biodiesel blends. In other words, similar fuel flow rates are supplied to the engine when operating with those blends. These 4 fuel flow rate conditions (M1 to M4, respectively) were investigated with the aim to examine the influence of engine load on engine combustion characteristics. M1 is close to the full load condition while M2, M3 and M4 are close to three-quarters, half and a quarter load conditions, respectively. Mode M2 has a similar fuel flow rate to the CN testing mode;

(6)   Operations of the engine under a constant input energy amount (J/min) conditions are also conducted. When operating an engine with different fuels having different heating values, the engine loading conditions (e.g., engine torque) are different from one fuel to the other. When comparing engine performance under the same conditions of engine load and speed, a constant input energy approach should be adopted, and this is one part of this study. Here, the investigation under constant input energy conditions limits only to CR = 15. Under this mode, a constant input energy amount of 463.6 J/min is applied for all fuel blends tested. This amount is corresponding to the energy of B0 at mode M1 (15.5 mL/min). This testing mode will be denoted as Q_{const} in this study.

3. Results and Discussion

3.1. Influence of Fuel Flow Rate Conditions

As mentioned, four different fuel flow rates have been tested in this study. Modes M1, M2, M3 and M4 note for 15.5, 13.0, 11.30 and 10.0 mL/min of fuel flow rates, respectively. These conditions may be close to full, three-quarters, half and a quarter engine load conditions. In-cylinder pressure signals outputted in this study are averaged from 50 consecutive engine cycles. Then, those averaged signals are used to compute the heat release rate (HRR), maximum in-cylinder pressure (p_{max}), rate of in-cylinder pressure rise (RPR) and MRPR. In this section, the influence of fuel flow rates on in-cylinder pressure and HRR will be reported. Discussions related to p_{max}, RPR and MRPR will be shown later on in the following sections.

Figure 2 shows an example of the in-cylinder pressure and HRR of B100 under the case of Q_{const} and t_{inj} = 16 DCA. Regarding the influence of fuel flow rates on in-cylinder pressure development and HRR, the trend in the in-cylinder pressure development and HRR is similar for all blends under a certain injection timing. Therefore, only an example of the in-cylinder pressure and HRR of B100 under t_{inj} = 16 DCA is reported in Figure 2. The full database may be provided upon request. It is quite clear from this figure that the fuel flow rates mainly affect the second combustion period named diffusion combustion. The influence of fuel flow rate on the first period (characterized by **SOC** and premixed combustion fraction) is minimal. Details about premixed and diffusion combustion will be discussed in the following paragraphs.

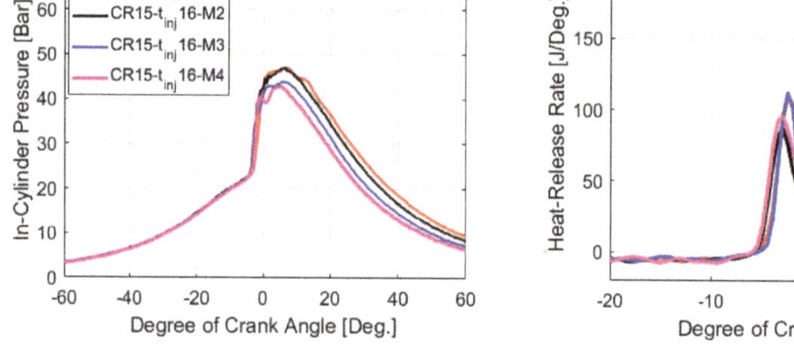

(**a**) P_{cyl}, B0, CR = 15, t_{inj} = 16 DCA BTDC.　　(**b**) HRR, B0, CR = 15, t_{inj} = 16 DCA BTDC.

Figure 2. Example of P_{cyl} (**a**) and heat release rate (HRR) (**b**) versus DCA (B0, at different modes M1 to M4, different fuel flow rates, t_{inj} = 16, CR15).

As can be seen from Figure 2a, the in-cylinder pressure signals obtained when operating the engine under different fuel flow rates (M1 to M4) are almost identical in the initial duration (before TDC in these cases). Then, a higher fuel flow rate leads to a higher in-cylinder pressure developed. Further combustion characteristics are detailed in Figure 2b, where the HRRs obtained under these fuel loading conditions are shown. Combustion of a premixed mixture is much faster with respect to diffusion combustion [36]. The premixed and diffusion combustion fractions are distinguished quite clearly through the HRR signals shown in Figure 2b. During the premixed combustion period, the HRR first rapidly rises, gets to its peak and significantly decreases (before TDC as shown in Figure 2b). Then, diffusion combustion takes place where the HRR is quite low with respect to that during premixed combustion.

It is quite clear from Figure 2b that the premixed combustions are identical amongst the fuel flow rates tested here while a higher fuel flow rate leads to a higher HRR during diffusion combustion. Qualitatively, loading conditions do not influence the **SOC** (where the HRR signals suddenly rise and become positive as shown in this figure. Since the evaporation and pre-mixing conditions including thermal conditions and pressure in the period from SOI to **SOC** are similar for all loading modes M1 to M4, the amount of fuel and air that is pre-mixed could be the same. This may be attributable to the similarity in ignition delay times observed in Figure 2b for all modes.

As mentioned earlier that the trends in in-cylinder pressure development and HRR, especially during premixed combustion, are identical amongst the fuel flow rates tested here, only mode M1 (closing to full load conditions) will be further investigated in the following sections. However, the full database may be made available upon request.

3.2. *Influence of Thermal Conditions at SOI*

Figure 3 shows the in-cylinder pressure signals for B0 and B100 under the injection mode M1 (15.5 mL/min used for all fuels), but under different injection timings (t_{inj} from 8 to 20 DCA BTDC). Figure 3a,b are for B0 at CR = 15 and B0 at CR = 17, respectively. Similarly, Figure 3c,d are for B100. Thermal conditions here mean the temperature and pressure at SOI depend on t_{inj} in the compression stroke. These figures shown here are examples to evaluate the influence of thermal conditions at SOI on in-cylinder pressure signals. In-cylinder pressure signals for other blends are not shown here as they show identical trends to these figures. It is simply noted that the in-cylinder pressure at CR = 15 (left figures, Figure 3a,c) is lower than that at CR = 17 (right figures, Figure 3b,d). Certainly, under higher CR, the engine creates higher in-cylinder temperature and pressure and this enhances the fuel evaporation, mixing and combustion. Engines with higher CR, therefore, normally show their higher thermal efficiency.

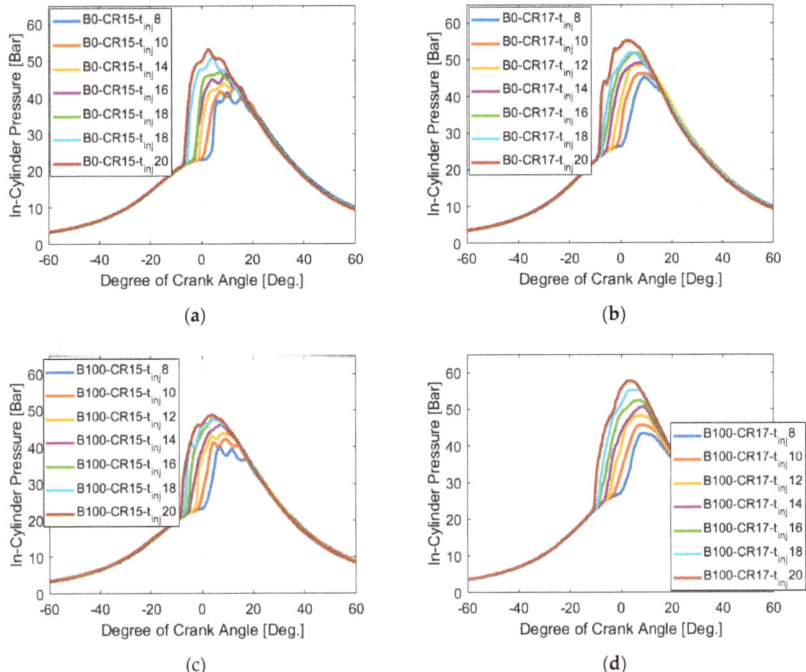

Figure 3. Influence of in-cylinder thermal conditions at SOI on the development of in-cylinder pressure. (**a**) B0, M1, compression ratio (CR) = 15; (**b**) B0, M1, CR = 17; (**c**) B100, M1, CR = 15; (**d**) B100, M1, CR = 17.

It is noted that the difference between p_{cylmax} shown in Figure 3a (B0 under CR = 15) and Figure 3b (B0 under CR = 17) is not that significant compared to the one observed in Figure 3c (B100 under CR = 15) and Figure 3d (B100 under CR = 17). This may be due to the improvement in the atomization, evaporation and combustion of B100 under high CR and long injection advanced timing conditions. By carefully observing, when t_{inj} increases over 14 DCA BTDC, p_{cylmax} developed under CR = 17 starts to rise more significantly compared to that under CR = 15. It is well known that biodiesel atomizes and evaporates poorer than fossil diesel. High-temperature conditions created under a high CR and injecting far from TDC (e.g., t_{inj} > 14 DCA in this case) could help biodiesel to enhance its atomization and evaporation significantly. The enhancement along with oxygen content in biodiesel may improve its combustion quality and this in turn causes the significant increase in p_{cylmax}. This phenomenon may be explained by carefully observing the combustion duration, position of ignition, position of peak pressure and combustion phasing, however, combustion is a very complex phenomenon and this is a subject for future study.

Along with HHV, viscosity, surface tension and cetane number are important parameters (even more important than HHV) impacting the fuel combustion characteristics and therefore in-cylinder pressure development. It is observed in this study that under some testing conditions like at CR = 17 and high advanced injection timing, B100 produces a higher maximum in-cylinder pressure with respect to that of B0. This might be attributed to the improvement in the atomization quality of B100 under these conditions along with the oxygen content in the fuel but this needs further investigations. Compared to conventional diesel, biodiesel has higher viscosity and surface tension and therefore poorer atomization quality [44]. One of the key characteristics of biodiesels is that the fuels contain oxygen in their molecules and this feature makes them special compared to fossil diesel. In compression ignition engines, the fuel oxygen enhances the combustion quality in the fuel spray's reaction zone, where the fuel–air mixture is rich (lack of oxygen) [35,43]. Under high advanced injection timings like t_{inj} = 18 and 20 DCA BTDC, biodiesel has a longer time to atomize and this along with the higher oxygen content of the fuel might enhance its combustion quality. This could be the reason for the higher cylinder pressure obtained here. Anyway, combustion is very complex and, as such, further investigations are required.

One of the most important parameters characterizing combustion is ignition delay and, as such, examining the delay time is critically important in studies related to engines and fuels. Basically, ignition delay is the period between the start of injection and start of combustion (SOC). Recording the start of injection can be straight forward through the engine management system (e.g., through the injection control system). Determining SOC, however, is quite challenging. In the literature, SOC or auto-ignition location is commonly defined as when the HRR locally becomes zero and reverses direction, although the natural flame emission was identified earlier than the SOC [35]. SOC is very sensitive to fuel molecule size and structure [43]. The first stage of combustion is called premixed burn duration. This duration is quite short [45] and leads to a high rate of heat release. The heat release rate during this stage is strongly dependent on the amount of air–fuel premixed during the ignition delay period [46]. The main combustion stage, diffusion combustion, is associated with a lower rate of heat release.

To investigate the influence of injection timing on combustion characteristics, HRR signals of B100 under mode M1 and different t_{inj} are shown in Figure 4. Qualitatively, ignition delay times could be identified in this figure where HRR locally becomes zero and reverses direction. One example of ignition delay time is indicated in Figure 4a for B100 under a condition of injection timing of 20 DCA. Quantitative information of ignition delay times will be shown later at the end of Section 3.3.

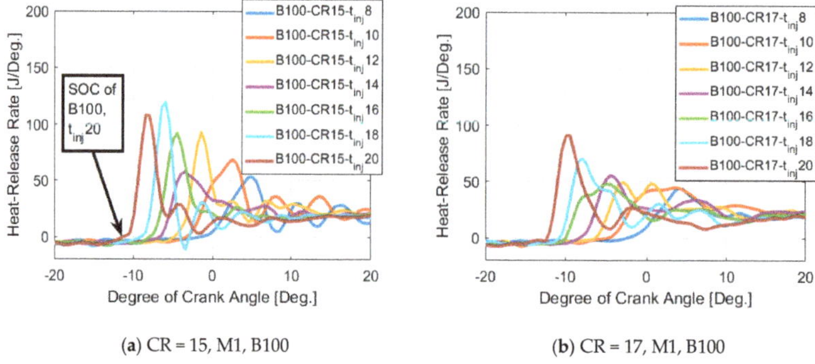

(**a**) CR = 15, M1, B100   (**b**) CR = 17, M1, B100

Figure 4. HRR of B100, CR = 15 (**a**) CR = 17 (**b**) at mode M1, different injection timings.

3.3. Influence of Blending Ratio

Figure 5a–f show in-cylinder pressure signals for all biodiesel blends (B0, B10, B20, B40, B60 and B100) under the injection mode M1. The left column (Figure 5a,c,e) is shown for CR = 15, while the right column (Figure 5b,d,f) is for CR = 17. The top row (Figure 5a,b) is for t_{inj} = 10 DCA BTDC, the middle row (Figure 5c,d) is for t_{inj} = 16 DCA BTDC and the bottom row (Figure 5e,f) is for t_{inj} = 20 DCA BTDC. Qualitatively, Figure 5 shows some differences in p_{max} developed by the engine when operating with those blends. Quantitative information about p_{max} developed by these biodiesel blends will be shown and discussed later in Section 3.6.

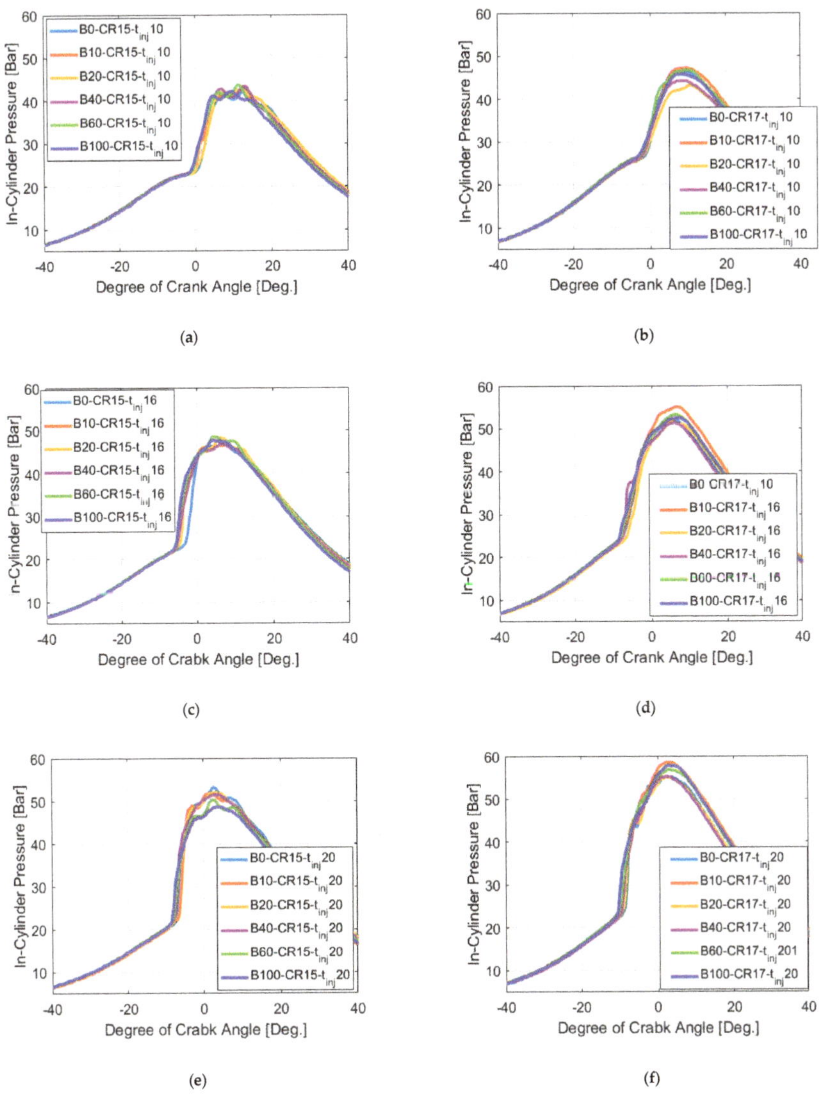

Figure 5. Influence of blending ratio on the development of in-cylinder pressure under mode M1. (**a**) M1, t_{inj} = 10, CR = 15; (**b**) M1, t_{inj} = 10, CR = 17; (**c**) M1, t_{inj} = 16, CR = 15; (**d**) M1, t_{inj} = 16, CR = 17; (**e**) M1, t_{inj} = 20, CR = 15; (**f**) M1, t_{inj} = 20, CR = 17.

It is noted here that the influence of blending ratio on the in-cylinder pressure shown in Figure 5a–f is not quite significant. However, in general, an increase in the blending ratio of biodiesel shifts the start of the in-cylinder pressure rise from the compression trace to the left. This could be an evidence of auto-ignitability enhancement when increasing the blending ratio and will be investigated further through examining the HRR, shown later in Figure 6. It is shown in Figure 5 that the difference in in-cylinder pressure is quite small. The thermal conditions at SOI may be on the side of high-temperature combustion (HTC) regimes but this needs to be further investigated. Nevertheless, it is found by Westbrook et al. in an earlier study [47] that, under low-temperature combustion (LTC) regimes, the influence of hydrocarbon fuels and biodiesels is observable. Under HTC, however, the influence is not quite as significant.

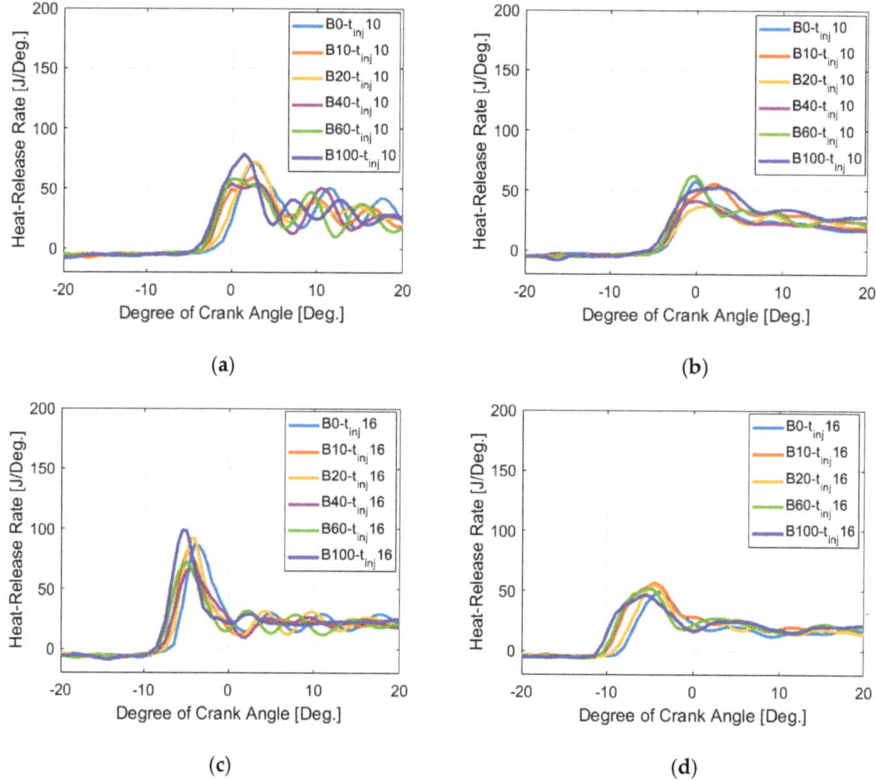

Figure 6. Influence of blending ratio on HRR under different constant fuel flow rate compression ratios. (**a**) CR = 15, M1, t_{inj} = 10 DCA BTDC; (**b**) CR = 17, M1, t_{inj} = 10 DCA BTDC; (**c**) CR = 15, M1, t_{inj} = 16 DCA BTDC; (**d**) CR17, M1, t_{inj} = 16 DCA BTDC.

Figure 6 shows the HRR of those blends tested in this study under mode M1. The left column (Figure 6a,c) is shown for CR = 15, while the right column (Figure 6b,d) is for CR = 17. The top row (Figure 6a,b) is for t_{inj} = 10 DCA BTDC and the bottom row (Figure 6c,d) is for t_{inj} = 20 DCA BTDC. It is quite clear here that adding biodiesel into the diesel–biodiesel blends leads to shorter ignition delay times. B100, generally, shows the shortest ignition delay, while B0 shows the longest ignition delay time. Varying the blending ratio from B0 to B100 leads to a difference of approximately 3 DCA in the ignition delay. This is obviously attributed to the higher CN of the biodiesel as reported earlier in this work. It is also noted that the differences amongst the HRRs at the start of combustion (SOC) are quite similar under CR = 15 (Figure 6a,c) and t_{inj} = 16 DCA BTDC, CR = 17 (Figure 6d).

Figure 6b shows the smallest gap amongst the HRRs at SOC compared to that shown in other figures here. Under the high CR and small t_{inj} reported in Figure 6b (CR = 17, t_{inj} = 10 DCA BTDC), the temperature at SOI is high and this is attributable to the small difference in HRR at the start of combustion observed in this case. Compared to the HRR obtained in Figure 6a,c for CR = 15, the HRR shown in Figure 6b,d for CR = 17 is lower. This is understandable as the higher in-cylinder temperature under higher CR causes a shorter ignition delay and smaller premixed combustion fraction.

Ignition delay times of B100 at mode M1 but different injection timings, corresponding to the HRR shown in Figure 4, are shown in Figure 7a. As can be seen from Figure 7a, the injection timings strongly affect the ignition delay, and this is due to the difference in the premixed combustion fraction observed earlier in Figure 4. In general, injecting fuel further from the top dead center (TDC) leads to a longer ignition delay and, as such, a higher premixed combustion fraction. Injecting the fuels close to (TDC), like the case of t_{inj} = 8 DCA, leads to a high temperature at SOI, Ti, and, as such, the fuel–air mixture is easier to be auto-ignited. Higher Ti leads to a shorter ignition delay time and smaller fraction of premixed combustion. Under CR = 17, Ti is higher and HRR is lower compared to those under CR = 15 (see Figure 4), and this is attributable to the shorter ignition delay under the higher CR conditions shown in Figure 7a.

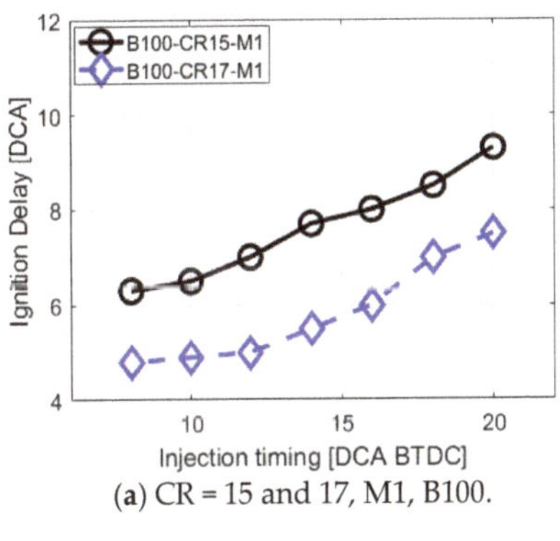

(**a**) CR = 15 and 17, M1, B100.

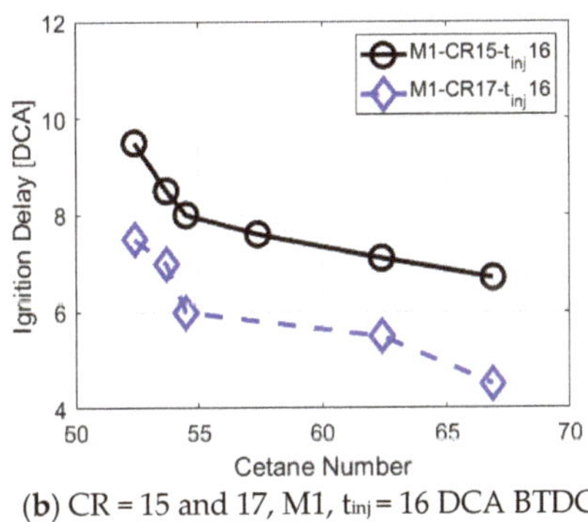

(**b**) CR = 15 and 17, M1, t_{inj} = 16 DCA BTDC, different blends.

Figure 7. Ignition delay of (**a**) B100 at mode M1 but different injection timings, corresponding to the HRR shown in Figure 4. (**b**) All blends at mode M1, t_{inj} = 16 DCA BTDC, corresponding to the HRR shown in Figure 6c,d.

Ignition delay times of all blends under mode M1 and injection timing t_{inj} = 16 DCA BTDC, corresponding to the HRR shown in Figure 6d,c for CR = 15 and CR = 17, respectively, are shown in Figure 7b. It is clear here that increasing the blending ratio from B0 to B100 leads to increasing the CN, and this is obviously attributed to decreasing the ignition delay times or improving the fuel blend reactivity. It is observed here again that the higher in-cylinder temperature under higher CR causes a shorter ignition delay and smaller premixed combustion fraction.

3.4. Development of In-Cylinder Pressure and HRR under Constant Input Energy Supplying Modes

In engine experiments, engines are controlled under two main conditions, namely engine speed and load. Speed here is the crankshaft rotations per minute (rpm), while engine load is determined through engine torque (N·m) or engine power (kW) [36]. To obtain similar engine torque, supplying constant input energy to the tested engine is normally adopted, regardless of the fuel used. This is to account for the difference in heating values amongst the fuels tested. It is noted here again that both methods of fuel supplied (equal volume flow rate [42,48] and equal input energy [21], respectively) are available in the literature and this sometime confuses the reader. On the one hand, the control system normally uses volume flow rate (through injection pressure and duration) to drive the injectors and, as such, the approach of using equal volume flow rate (L/min) supplied for the engine when testing different fuels is quite common in the literature. This is particularly true with engines equipped with mechanical injection systems such as the one used in [48] for controlling the mechanical system to supply constant input energy, which is quite challenging. On the other hand, with different heating value fuels like biodiesel blends tested in this study, they produce different engine torque or different engine loading conditions when supplying an equal volume flow rate for all fuels. Relative comparisons of the engine performance when operating with different fuels under different engine loads, as such, may not be meaningful [9].

Figure 8a,d show in-cylinder pressure signals for all biodiesel blends (B0, B10, B20, B40, B60 and B100) under the injection mode Q_{const} (constant input energy (J/min) supplied amongst the biodiesel blends). For all blends, an energy flow rate of 463.6 J/min is supplied to the engine under this mode. This amount corresponds to the fuel volume flow rate of B0 at mode M1 (15.5 mL/min). The fuel volume flow rates of other blends can be calculated by multiplying 463.6 J/min with the blends' heating value provided in Table 2. This mode was conducted only for CR = 15 and Figure 8 shows the pressure signals developed under t_{inj} = 10 (Figure 8a), t_{inj} = 12 (Figure 8b), t_{inj} = 16 (Figure 8c) and t_{inj} = 20 (Figure 8d). It is clear that the influence of fuel properties can be ignored here.

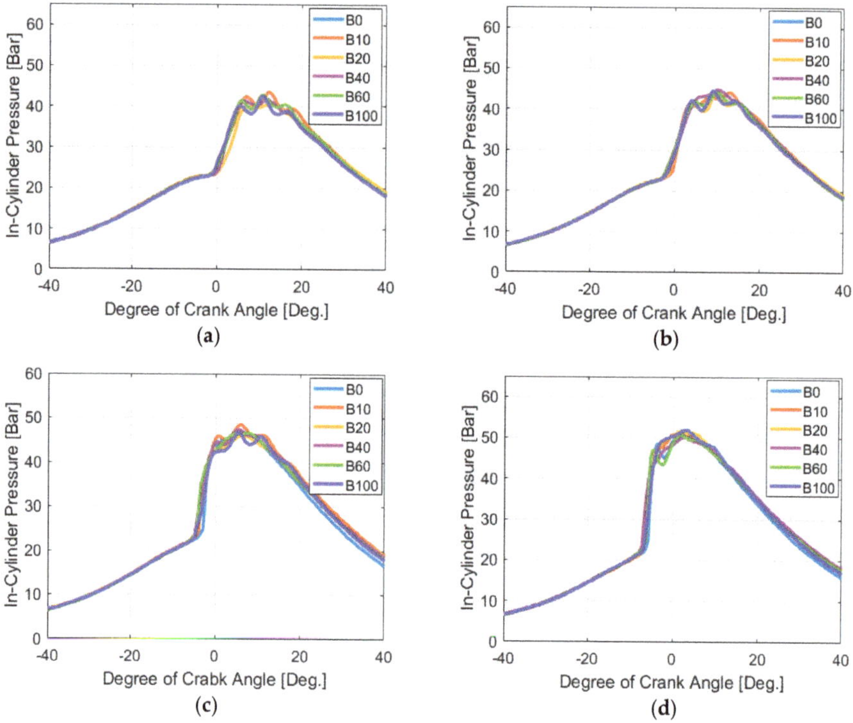

Figure 8. Influence of blending ratio on in-cylinder pressure development when supplying constant input energy amounts (Q_{const}) to the cylinder. (**a**) Q_{const}, CR = 15, t_{inj} = 10 DCA BTDC; (**b**) Q_{const}, CR = 15, t_{inj} = 12 DCA BTDC; (**c**) Q_{const}, CR = 15, t_{inj} = 16 DCA BTDC; (**d**) Q_{const}, CR = 15, t_{inj} = 20 DCA BTDC.

Nevertheless, results observed in Figure 8a,d indicate that the influence of blending ratios on the **SOC** is quite similar to the results observed earlier in Figure 5 showing results under mode M1, with a constant fuel volume flow rate. Figure 8 shows that although using the equal input energy approach brings the engine loads close amongst the blends tested, in this special equipment (the CFR engine), the auto-ignitability is mainly driven by the chemical profile of the testing fuels rather than the engine loads, and this is in a good agreement with the discussion shown earlier in Section 3.1. Again, Section 3.1 reports that the influence of engine load on in-cylinder development around the **SOC** is minimal. This may be probably true only in the CFR engine, as in this tool, the intake air temperature is well controlled and kept constant at 65 °C for all testing conditions. Under a similar thermal condition at SOI, the fuel evaporation rate and pre-mixing during the ignition delay period could be similar under different engine loading conditions and, as such, the auto-ignitability mainly depends on the fuel reactivity.

Figure 9 shows the HRR of the biodiesel blends under constant input energy modes, at injection timings t_{inj} = 10 DCA (Figure 9a) and t_{inj} = 16 DCA (Figure 9b). The influence of blending ratio on the ignition delay and premixed combustion fraction is quite clear and this is also similar to what has been observed in Figure 6 for similar volume flow rate conditions. A higher biodiesel fraction in the blend leads to a shorter ignition delay and this is attributed to the higher CN of the biodiesel.

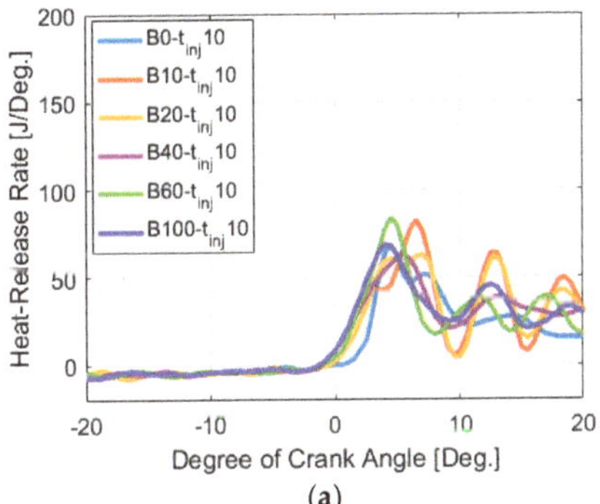
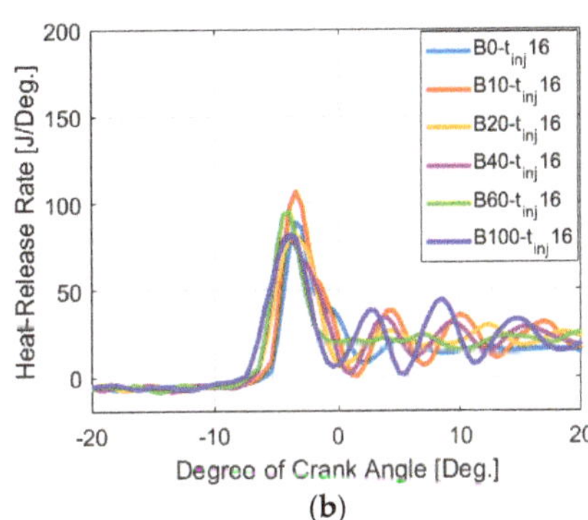

(a) (b)

Figure 9. HRR of biodiesel blends when supplying constant input energy amounts (Q_{const}) to the cylinder. (a) Q_{const}, CR = 15, t_{inj} = 10 DCA BTDC; (b) Q_{const}, CR15, t_{inj} = 16 DCA BTDC.

3.5. Rate of In-Cylinder Pressure Rise

Figure 10a,d show the RPR for all biodiesel blends (B0, B10, B20, B40, B60 and B100) under the injection mode M1. The left column (Figure 10a,c) is shown for CR = 15, while the right one (Figure 10b,d) is for CR = 17. The top row (Figure 10a,b) is for t_{inj} = 8 DCA BTDC, while the bottom row (Figure 10c,d) is for t_{inj} = 20 DCA BTDC.

It is quite interesting from these figures that under high CR and/or short advanced injection timing conditions, like the cases shown in Figure 10a (t_{inj} = 8 DCA BTDC, CR15) and Figure 10b (t_{inj} = 8 DCA BTDC, CR17), the influence of fuel blending ratio on RPR is almost ignorable. Again, the combustion happening in these cases could fall into HTC regimes as mentioned earlier and, as such, the influence of fuel properties on ignition delay is minimal. It is also noted here that the HRR under CR = 17 is lower compared to that under CR = 15. Under higher CR conditions, the ignition delay time is shorter due to the higher in-cylinder temperature and pressure at the end of the compression stroke. The shorter ignition delay leads to a higher amount of air–fuel premixed.

Under low CR and/or long advanced injection timing conditions, like the cases shown in Figure 10c (t_{inj} = 20 DCA BTDC, CR15) and Figure 10d (t_{inj} = 20 DCA BTDC, CR17), however, the blending ratios significantly affect the RPR. A general trend observed in Figure 10c,d is that the fuels with lower CN will have higher RPR and this is due to their longer ignition delay and therefore high premixed combustion fraction as discussed briefly earlier in this study. For example, B100 (highest CN) shows the lowest RPR and this rapid rise in the RPR of B100 occurs earlier compared to other blends, as shown in Figure 10c,d.

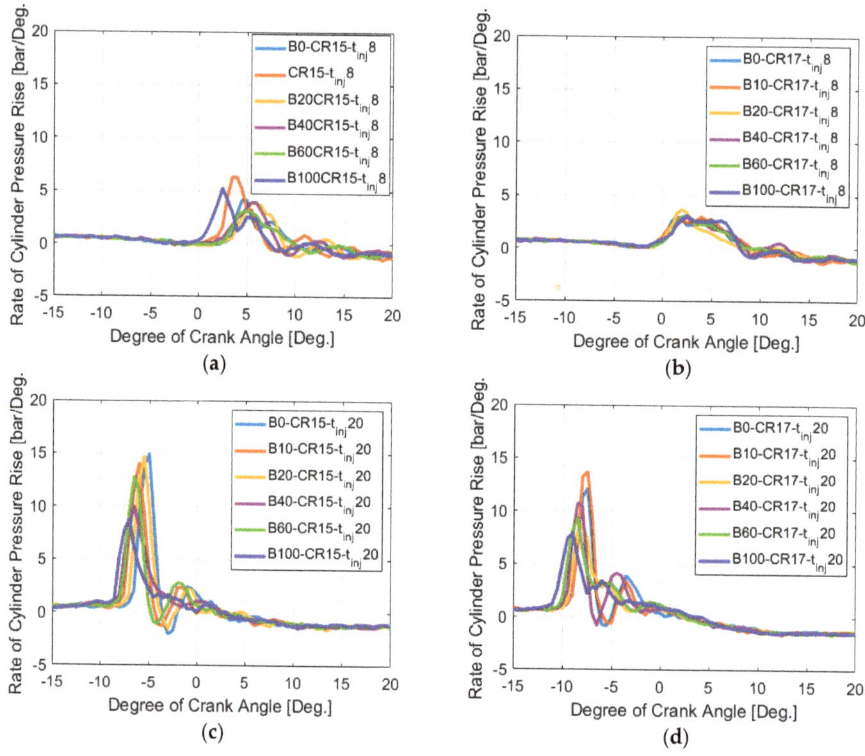

Figure 10. Rate of in-cylinder pressure rise versus DCA under different operating conditions. (**a**) M1, t_{inj} = 8 TDC BTDC, CR = 15; (**b**) M1, t_{inj} = 8 DCA BTDC, CR = 17; (**c**) M1, t_{inj} = 20 TDC BTDC, CR = 15; (**d**) M1, t_{inj} = 20 TDC BTDC, CR = 17.

3.6. Maximum In-Cylinder Pressure

Figure 11a,b show the maximum value of in-cylinder pressure, p_{cylmax}, versus diesel–biodiesel blending ratio. Results shown in Figure 11a are for CR = 15 under the injection mode Q_{const}, while Figure 11b is for CR = 17 under the injection mode M1. As can be seen from Figure 11a,b, the influence of blending ratio on p_{cylmax} is not significant except for the case of B10. A general trend observed for B10 in these figures is that when increasing the blending ratio from 0 to 10%, p_{cylmax} generally increases. The higher p_{cylmax} of B10 compared to B0 and B20 could be due to the lubricant enhancement when operating the engine with low blending ratios of biodiesel–diesel mixture. It has been claimed in [1,49] that adding a small amount of biodiesel into diesel fuel (e.g., 2–10 vol%) will help to improve the engine lubricant and therefore the thermal efficiency. The improvement of the lubricant is achieved due to the high viscosity of biodiesel compared to fossil diesel as can be seen in Table 2. When lubricant conditions are improved, the piston–cylinder thermal gap will be decreased as this may be attributed to the increase in the p_{cylmax} observed for B10 in this case. When the blending ratio is high enough, like B20 in this study, the enhancement will not be achieved. Biodiesels are found to have higher lubricity with respect to fossil diesel [9,44]. However, biodiesels can contribute to the formation of deposits [9] and their higher viscosity and surface tension lead to their poorer atomization and vaporization [44]. The above-mentioned factors could be attributed to impairing the lubricant benefit when utilizing high-blending ratio diesel–biodiesel mixtures like B20 tested here.

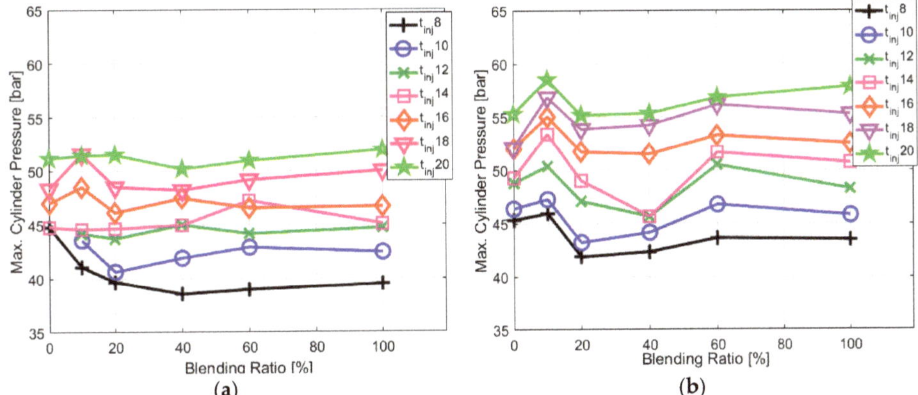

Figure 11. Maximum of in-cylinder pressure versus cetane number (CN) (each curve shown here contains data of all biodiesel blends tested). (**a**) Q_{const}, CR = 15; (**b**) M1, CR = 17.

It is also clear from Figure 11 that injecting fuel closer to the TDC leads to a lower p_{cylmax}. The temperature at SOI is higher when injecting fuel closer to the TDC and this will lead to a shorter ignition delay and smaller fraction of premixed combustion. The smaller premixed combustion fraction is the main reason resulting in the lower p_{cylmax}. It is noted here that this trend was qualitatively observed earlier in Figures 3 and 4 and the quantitative result is reported here.

3.7. Maximum Rate of In-Cylinder Pressure Rise

Figure 12a,b show the MRPR for all biodiesel blends (B0, B10, B20, B40, B60 and B100) versus the thermal condition at SOI, 1000/Ti. Figure 12a is shown for CR = 15, while Figure 12b is for CR = 17. It can be seen from these figures that the MRPR is significantly affected by the thermal condition at SOI, thus a lower injection temperature (towards to the right side of the 1000/Ti axis) leads to a higher MRPR. It was noted earlier that low temperature at SOI leads to a long ignition delay and high premixed combustion fraction, and this is attributed to the high MRPR. Furthermore, MRPR values of those biodiesel blends are quite diverse, except for CR = 17, and in the range of 1000/Ti smaller than 1.23 shown in Figure 12b. Although it is not quite consistent, increasing the blending ratio generally deceases the MRPR. Under CR = 17 and in the range of 1000/Ti smaller than 1.23 shown in Figure 12b, the diversion of the MRPR amongst the blends disappears. Under the high CR and higher injection temperature, the combustion here may fall right in the low temperature range of the HTC strategy. It was observed earlier [47] that combustion characteristics of biodiesels and diesel are identical under HTC conditions.

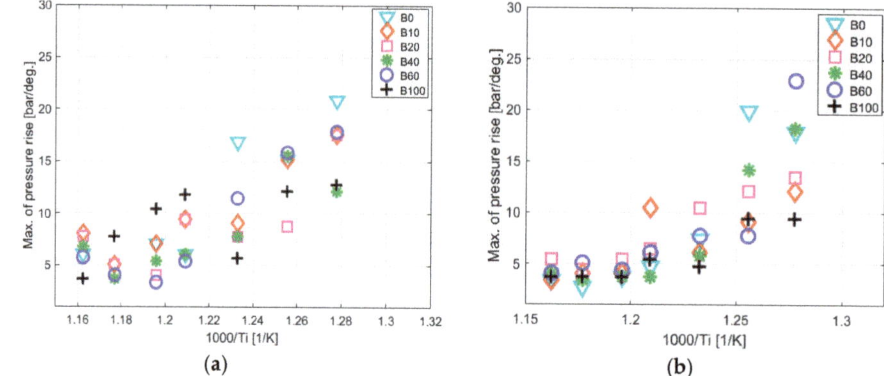

Figure 12. Max rate of in-cylinder pressure rise versus 1/Ti. (**a**) M5, CR15; (**b**) M5, CR17.

4. Conclusions

An analysis of biodiesel blend combustion characteristics under a wide range of thermal conditions of a CFR engine has been extensively carried out in this study. It is observed that the oxygen content in biodiesel has significant effects on the fuel auto-ignitability. A higher blending ratio of a biodiesel–diesel mixture leads to its higher CN. The cetane number of pure biodiesel (B100) is almost 30% higher than that of fossil diesel (B0), and this could be due to the oxygen content in the biofuel. In this CFR-F5 engine, it is observed that varying the engine load has minimal effect on the premixed combustion. The HRR observed during the premixed combustion period is identical when testing the engine under different fuel flow rates. Higher fuel flow rates, however, lead to a higher HRR during diffusion combustion.

A higher temperature leads to a smaller premixed combustion fraction (due to a shorter ignition delay) and therefore a higher MRPR. At the same thermal conditions at SOI, when operating the engine under CR = 15 and 17, increasing the blending ratio generally has quite a small effect on in-cylinder pressure development, except for B10. The higher p_{cylmax} of B10 observed here compared to that of B0 and B20 could be due to the lubricant enhancement when operating the engine with low blending ratios of biodiesel–diesel mixture. When using a low-blending ratio mixture like B10 in this case, the higher viscosity of the biodiesel may help to lower the thermal gap between the piston and the cylinder, and this may lead to an increase in in-cylinder pressure. Under high blending ratios, this benefit is not achieved.

Author Contributions: Conceptualization, methodology, validation, analysis, investigation, and resources: V.H.N., P.X.P., K.T.N., T.V.P. and M.Q.D.; Writing—original draft preparation: P.X.P.; writing—review and editing: P.X.P., V.H.N., M.Q.D., K.T.N. and T.V.P. Supervision: V.H.N., P.X.P. and K.T.N.; All authors have read and agreed to the published version of the manuscript.

Funding: This research received no external funding.

Conflicts of Interest: The authors declare no conflict of interest.

References

1. Krahl, J.; Knothe, G.; Van Gerpen, J.H. *Biodiesel Handbook*; AOCS Press: Urbana, IL, USA, 2010.
2. Kegl, B.; Kegl, M.; Pehan, S. Green diesel engines. In *Biodiesel Usage in Diesel Engines*; Springer: London, UK, 2013.
3. Demirbas, A. Production of biodiesel from algae oils. *Energy Sources Part A Recovery Util. Environ. Eff.* **2008**, *31*, 163–168. [CrossRef]
4. Rahman, M.; Pourkhesalian, A.; Jahirul, M.; Stevanovic, S.; Pham, P.; Wang, H.; Masri, A.; Brown, R.J.; Ristovski, Z. Particle emissions from biodiesels with different physical properties and chemical composition. *Fuel* **2014**, *134*, 201–208. [CrossRef]

5. Pham, P.; Bodisco, T.A.; Stevanovic, S.; Rahman, M.; Wang, H.; Ristovski, Z.; Brown, R.; Masri, A. Engine performance characteristics for biodiesels of different degrees of saturation and carbon chain lengths. *SAE Int. J. Fuels Lubr.* **2013**, *6*, 188–198. [CrossRef]

6. Pourkhesalian, A.; Stevanovic, S.; Salimi, F.; Rahman, M.; Wang, H.; Pham, P.; Bottle, S.; Masri, A.; Brown, R.; Ristovski, Z. Influence of fuel molecular structure on the volatility and oxidative potential of biodiesel particulate matter. *Environ. Sci. Technol.* **2014**, *48*, 12577–12585. [CrossRef]

7. Nguyen, V.H.; Pham, P.X. Biodiesels: Oxidizing enhancers to improve CI engine performance and emission quality. *Fuel* **2015**, *154*, 293–300. [CrossRef]

8. Vargas, A.P.F.; Delgado, R.; Hernández, E.; Suástegui, J.A. Performance Analysis of a Compression Ignition Engine Using Mixture Biodiesel Palm and Diesel. *Sustainability* **2019**, *11*, 4918. [CrossRef]

9. Lapuerta, M.; Armas, O.; Rodriguez-Fernandez, J. Effect of biodiesel fuels on diesel engine emissions. *Prog. Energy Combust. Sci.* **2008**, *34*, 198–223. [CrossRef]

10. Hasan, M.; Rahman, M. Performance and emission characteristics of biodiesel–diesel blend and environmental and economic impacts of biodiesel production: A review. *Renew. Sustain. Energy Rev.* **2017**, *74*, 938–948. [CrossRef]

11. Damanik, N.; Ong, H.C.; Tong, C.W.; Mahlia, T.M.I.; Silitonga, A.S. A review on the engine performance and exhaust emission characteristics of diesel engines fueled with biodiesel blends. *Environ. Sci. Pollut. Res.* **2018**, *25*, 15307–15325. [CrossRef]

12. Lai, J.Y.; Lin, K.C.; Violi, A. Biodiesel combustion: Advances in chemical kinetic modeling. *Prog. Energy Combust. Sci.* **2011**, *37*, 1–14. [CrossRef]

13. Assanis, D.N.; Filipi, Z.S.; Fiveland, S.B.; Syrimis, M. A predictive ignition delay correlation under steady-state and transient operation of a direct injection diesel engine. *J. Eng. Gas Turbines Power* **2003**, *125*, 450–457. [CrossRef]

14. Shahabuddin, M.; Liaquat, A.; Masjuki, H.; Kalam, M.; Mofijur, M. Ignition delay, combustion and emission characteristics of diesel engine fueled with biodiesel. *Renew. Sustain. Energy Rev.* **2013**, *21*, 623–632. [CrossRef]

15. Singh, G.; Pham, P.; Kourmatzis, A.; Masri, A. Effect of electric charge and temperature on the near-field atomization of diesel and biodiesel. *Fuel* **2019**, *241*, 941–953. [CrossRef]

16. Pham, P.; Kourmatzis, A.; Masri, A. Local characteristics of fragments in atomizing sprays. *Exp. Therm. Fluid Sci.* **2018**, *95*, 44–51. [CrossRef]

17. Campbell, M.F.; Davidson, D.F.; Hanson, R.K.; Westbrook, C.K. Ignition delay times of methyl oleate and methyl linoleate behind reflected shock waves. *Proc. Combust. Inst.* **2013**, *34*, 419–425. [CrossRef]

18. Jaichandar, S.; Kumar, P.S.; Annamalai, K. Combined effect of injection timing and combustion chamber geometry on the performance of a biodiesel fueled diesel engine. *Energy* **2012**, *47*, 388–394. [CrossRef]

19. Dornelles, H.; Antolini, J.; Sari, R.; Nora, M.D.; Machado, P.R.; Martins, M. Analysis of engine performance and combustion characteristics of diesel and biodiesel blends in a compression ignition engine. *SAE Tech. Pap.* **2016**. [CrossRef]

20. Sayin, C.; Gumus, M.; Canakci, M. Effect of fuel injection pressure on the injection, combustion and performance characteristics of a DI diesel engine fueled with canola oil methyl esters-diesel fuel blends. *Biomass Bioenergy* **2012**, *46*, 435–446. [CrossRef]

21. Zhang, R.; Pham, P.X.; Kook, S.; Masri, A.R. Influence of biodiesel carbon chain length on in-cylinder soot processes in a small bore optical diesel engine. *Fuel* **2019**, *235*, 1184–1194. [CrossRef]

22. Duong, M.Q.; Nguyen, V.H.; Pham, P.X. Development of Empirical Correlations for Ignition Delay in a Single Cylinder Engine Fueled with Diesel/Biodiesel Blends. In Proceedings of the 2019 International Conference on System Science and Engineering (ICSSE), Dong Hoi, Vietnam, 20–21 July 2019; pp. 614–618.

23. Rothamer, D.A.; Murphy, L. Systematic study of ignition delay for jet fuels and diesel fuel in a heavy-duty diesel engine. *Proc. Combust. Inst.* **2013**, *34*, 3021–3029. [CrossRef]

24. Pham, P.X.; Kourmatzis, A.; Masri, A.R. Spray characterization of ethanol and saturated biodiesels of different carbon chain lengths. In Proceedings of the Australian Combustion Symposium, Perth, WA, Australia, 6–8 November 2013; pp. 392–395.

25. Kourmatzis, A.; Pham, P.X.; Masri, A.R. Air assisted atomization and spray density characterization of ethanol and a range of biodiesels. *Fuel* **2013**, *108*, 758–770. [CrossRef]

26. Wei, M.; Gao, Y.; Yan, F.; Chen, L.; Feng, L.; Li, G.; Zhang, C. Experimental study of cavitation formation and primary breakup for a biodiesel surrogate fuel (methyl butanoate) using transparent nozzle. *Fuel* **2017**, *203*, 690–699. [CrossRef]

27. Pham, P.; Kourmatzis, A.; Masri, A. Secondary atomization characteristics of biofuels with different physical properties. In Proceedings of the 19th Australasian Fluid Mechanics Conference (AFMC), Melbourne, Australia, 8–11 December 2014; pp. 1–4.

28. Kourmatzis, A.; Pham, P.X.; Masri, A.R. Characterization of atomization and combustion in moderately dense turbulent spray flames. *Combust. Flame* **2015**, *162*, 978–996. [CrossRef]

29. Haik, Y.; Selim, M.Y.; Abdulrehman, T. Combustion of algae oil methyl ester in an indirect injection diesel engine. *Energy* **2011**, *36*, 1827–1835. [CrossRef]

30. Le Breton, M.D. A Repeatability Test on the CFR Cetane Engine. *SAE Tech. Pap.* **1984**. [CrossRef]

31. Singh, B.; Shukla, S. Experimental analysis of combustion characteristics on a variable compression ratio engine fuelled with biodiesel (castor oil) and diesel blends. *Biofuels* **2016**, *7*, 471–477. [CrossRef]

32. Dash, S.K.; Lingfa, P.; Chavan, S.B. Combustion analysis of a single cylinder variable compression ratio small size agricultural DI diesel engine run by Nahar biodiesel and its diesel blends. *Energy Sources Part A Recovery Util. Environ. Eff.* **2020**, *42*, 1681–1690. [CrossRef]

33. Tesfa, B.; Mishra, R.; Zhang, C.; Gu, F.; Ball, A. Combustion and performance characteristics of CI (compression ignition) engine running with biodiesel. *Energy* **2013**, *51*, 101–115. [CrossRef]

34. Pham, P.X.; Bodisco, T.A.; Ristovski, Z.D.; Brown, R.J.; Masri, A.R. The influence of fatty acid methyl ester profiles on inter-cycle variability in a heavy duty compression ignition engine. *Fuel* **2014**, *116*, 140–150. [CrossRef]

35. Dec, J.E. A conceptual model of DL diesel combustion based on laser-sheet imaging. *SAE Trans.* **1997**, *106*, 1319–1348.

36. Heywood, J. *Internal Combustion Engine Fundamentals*, 2nd ed.; McGraw-Hill Education: New York, NY, USA, 2018.

37. Hasegawa, M.; Shimasaki, Y.; Yamaguchi, S.; Kobayashi, M.; Sakamoto, H.; Kitayama, N.; Kanda, T. Study on Ignition Timing Control for Diesel Engines Using In-Cylinder Pressure Sensor. *SAE Tech. Pap.* **2006**. [CrossRef]

38. Desantes, J.M.; Galindo, J.; Guardiola, C.; Dolz, V. Air mass flow estimation in turbocharged diesel engines from in-cylinder pressure measurement. *Exp. Therm. Fluid Sci.* **2010**, *34*, 37–47. [CrossRef]

39. Lapuerta, M.; Armas, O.; Hernández, J. Diagnosis of DI Diesel combustion from in-cylinder pressure signal by estimation of mean thermodynamic properties of the gas. *Appl. Therm. Eng.* **1999**, *19*, 513–529. [CrossRef]

40. Payri, F.; Broatch, A.; Tormos, B.; Marant, V. New methodology for in-cylinder pressure analysis in direct injection diesel engines—application to combustion noise. *Meas. Sci. Technol.* **2005**, *16*, 540. [CrossRef]

41. ASTM. *Standard Test Method for Cetane Number of Diesel Fuel Oil*; ASTM International: West Conshohocken, PA, USA, 2010.

42. Pham, P.X.; Nguyen, V.H. Residue-Based Biodiesel: Experimental Investigation into Engine Combustion and Emission Formation. *J. Energy Eng.* **2017**, *143*, 04017047. [CrossRef]

43. Westbrook, C.K. Biofuels combustion. *Annu. Rev. Phys. Chem.* **2013**, *64*, 201–219. [CrossRef] [PubMed]

44. Pandey, R.K.; Rehman, A.; Sarviya, R. Impact of alternative fuel properties on fuel spray behavior and atomization. *Renew. Sustain. Energy Rev.* **2012**, *16*, 1762–1778. [CrossRef]

45. Heywood, J.B. *Internal Combustion engine fundamentals*, 1a Edição; McGraw-Hill Education: NewYork, NY, USA, 1988.

46. Kamimoto, T.; Kobayashi, H. Combustion processes in diesel engines. *Prog. Energy Combust. Sci.* **1991**, *17*, 163–189. [CrossRef]

47. Westbrook, C.K.; Naik, C.V.; Herbinet, O.; Pitz, W.J.; Mehl, M.; Sarathy, S.M.; Curran, H.J. Detailed chemical kinetic reaction mechanisms for soy and rapeseed biodiesel fuels. *Combust. Flame* **2011**, *158*, 742–755. [CrossRef]

48. Manigandan, S.; Gunasekar, P.; Devipriya, J.; Nithya, S. Emission and injection characteristics of corn biodiesel blends in diesel engine. *Fuel* **2019**, *235*, 723–735. [CrossRef]

49. Tomic, M.; Savin, L.; Micic, R.; Simikic, M.; Furman, T. Possibility of using biodiesel from sunflower oil as an additive for the improvement of lubrication properties of low-sulfur diesel fuel. *Energy* **2014**, *65*, 101–108. [CrossRef]

 *sustainability* 

Article

The Impact of Fractional Composition on the Mechanical Properties of Agglomerated Logging Residues

Tomasz Nurek [1,*], Arkadiusz Gendek [1], Kamil Roman [2] and Magdalena Dąbrowska [1]

[1]   Department of Biosystems Engineering, Institute of Mechanical Engineering,
Warsaw University of Life Sciences—SGGW, Nowoursynowska 164, 02-787 Warsaw, Poland;
arkadiusz_gendek@sggw.edu.pl (A.G.); magdalena_dabrowska@sggw.edu.pl (M.D.)

[2]   Department of Technology and Entrepreneurship in Wood Industry,
Institute of Wood Sciences and Furniture, Warsaw University of Life Sciences—SGGW,
Nowoursynowska 159, 02-787 Warsaw, Poland; kamil_roman@sggw.edu.pl

*   Correspondence: tomasz_nurek@sggw.edu.pl; Tel.: +48-22-59-345-16

Received: 18 June 2020; Accepted: 28 July 2020; Published: 29 July 2020

Abstract: Fractional composition, as well as the temperature of the agglomeration process, affect the quality and mechanical properties of briquettes. In this research, shredded forest logging residues were investigated. Compaction tests were carried out for several specially prepared mixtures made of shares of fractions with different particle sizes. The moisture content, density of briquettes, specific work of compaction, mechanical durability, and biomass susceptibility to compaction were analyzed. Studies have confirmed the significant impact of the fractional composition of compacted biomass on its susceptibility to process parameters and the quality of the final product. Statistical analysis confirmed that the density of the briquette, its durability, the specific work of compaction, and the susceptibility of the tested biomass to compaction strongly depend on the particle size of the compacted biomass. An increase in temperature to 73 °C increased specific work by 40% and contributed to the high quality of briquettes in the range from 0.768 to 1.14 g·cm^{-3}.

Keywords: durability; mechanical parameters; briquettes; agglomeration; fractional composition

---

1. Introduction

The increase in human development causes an increase in energy demand in the modern world. This can inflict many negative effects of energy production methods on the environment. The increasing consumption of fossil fuels from year to year has resulted in the emission of huge amounts of greenhouse gases and pollutants into the atmosphere. In the longer term, this will lead to the further deterioration of the natural environment [1]. The protection of pure air and reduction of gases emitted to the atmosphere are very important for societies, not only of the European Union [2,3] but of all countries in the world. To promote sustainable development, support of renewable energy generation projects such as biomass, and promoting the transformation of the energy structure towards the diversification of these sources is crucial.

Faced with pressures related to the environmental degradation and scarcity of energy resources, demand for renewable energy with low environmental pollution is growing more and more, therefore it is extremely important to use sound and scientific decision-making methods regarding material processing and biomass energy production. The increase in demand for replacement of conventional fuels and reduction of CO_2 emissions motivates the search for new products based on green energy [4,5]. One of the raw materials with huge potential are forest residues. Therefore, the authors decided to conduct detailed research on this not-well-known renewable material.

There are many publications that specify the guidelines for the harvesting and processing of biomass, and also determine how energy from biomass may affect the environment. The research has shown that only a few of the ways in which biomass is processed relate to the processing of biomass to bioenergy [6]. Aherne et al. [7] found that only harvesting of aboveground wood biomass, i.e., crowns and branches, would have an indifferent effect on the environment and would not deplete the cationic base resources of the soil in the long term. On the other hand, Eggers et al. [8] have pointed out that the development of the innovative and cost-effective management systems for harvesting biomass from the young stands can provide the possibility of supplying a significant amount of the bioenergy while preserving the biodiversity or the ecosystem value.

The logging residues are un-merchantable wooden by-products created in the process of obtaining wood from forests and in the process of treating forest stands while cutting the undergrowth, which can be an important resource in energetic biomass. Due to their dispersion over large forest areas, they are difficult to collect, which results in high costs of obtaining them for e.g., fragmentation and agglomeration [9]. In addition, the costs of transporting such material and its storage are also high in terms of low bulk density, low energy content per volume, and high moisture content of the material [10]. Biomass briquetting can reduce these costs, emissions of gases, and the risk of fires [11]. Precisely for fire-fighting and breeding reasons, logging residues cannot remain on the forest surface and must be cleared before performing subsequent treatments. They can be used for energy purposes in their original form, after processing into bundles [12,13] or chips [14], where the chipping precursors were the Scandinavian countries [15]. Due to the fact that logging residues as energy biomass are not yet a fully utilized energy source, they could constitute, after fragmentation, a suitable feedstock for refined fuels—i.e., briquettes production. Considering the possibility of producing briquettes from shredded logging residues, their heterogeneous composition should be taken into account. In contrast to the compaction of wood chips, sawdust, or other plant waste, logging residues contain wood, bark, leaves, needles, and mineral contaminants from forest soil.

Plant waste from the wood industry with uniform composition and homogeneous particle size (sawdust, shavings, wood dust) has been widely used for briquette production for many years. Researchers working on this type of biomass examined the physical parameters (bulk density, moisture content, and particle size), chemical parameters (carbon, hydrogen, nitrogen, sulfur, oxygen and ash contents) as well as energy parameters (net and gross calorific values) of biomass intended for energy purposes. Many scientists have already dealt with these problems, in relation to logging residues [16]. However, there are no descriptions in the literature regarding the production of briquettes from contaminated logging residues with heterogeneous composition and heterogeneous particle size [11].

Knowledge of particle sizes is crucial in terms of briquettes quality, and particle sizes influence the mechanical properties of agglomerates. In the case of shredding the logging residues, the particle size distribution [17] and the size of the obtained chips mainly depend on the size of the wood chipper [18], wood species [19], the part of the tree, or the setting and sharpening of the knife [20]. Cutting with a blunt knife gives chips with smaller dimensions and thus increases the proportion of fine fractions [21,22].

In many publications, the authors emphasize that in order to obtain a briquette with an adequate durability described by the durability coefficient [23], it is necessary to choose not only the appropriate process parameters (compaction pressure, temperature, length-to-diameter ratio), but also biomass parameters (moisture content, particle size and distribution, lignin content, type of material, etc.) [24–28].

According to the results of tests on briquettes made of different particle sizes (2–5 mm and 7–10 mm), which were conducted by Gürdil and Demirel [29], it was found that briquettes made of particles of smaller sizes had smoother surfaces than those produced from larger particles, were more durable, and were characterized by higher densities, which was also confirmed by the studies of other authors [25,30–33].

Another parameter affecting the quality of briquettes is temperature. According to Taulbee [31], increasing the compaction temperature increases the strength of briquettes. Pressure and temperature are important parameters during materials compaction and affect the optimization of the process [34]. In addition, the increase in temperature causes plasticizing of the particles and activation of natural binders in the material [35].

Therefore, the aim of this study was to investigate and determine the effect of the fractional composition of shredded forest logging residues on the mechanical properties of briquettes obtained from them in the pressure agglomeration process.

2. Materials and Methods

2.1. Material

The investigated material was obtained from the forest area in the Chojnów forest district in Poland (GPS WGS84: 52.0492 N; 21.0563 E). The mixture of logging residues, including branches and needles from 80-year-old Scots pine (*Pinus sylvestris* L.) wood, was processed in a beater shredder BT 13HP-90 mm (REDMET, Dębica, Poland). The detailed characteristics of the raw material used for the briquetting was described in previous publications [36,37].

Shredded logging residues were divided into fractions using a sieve separator, according to the ISO 17827-1 [38] and ISO 17827-2 standards [39]. Four fraction groups were separated during the separation. During the 120-s trial, the research material was divided into the following fractions: f_1—(0 ÷ 1 mm), f_2—(1 ÷ 4 mm), f_3—(4 ÷ 8 mm), f_4—(8 ÷ 16 mm). The separated fractions were weighed. The mass of the i-th fraction (m_{fi}) was determined using an electronic scale with an accuracy of ±0.01 g. The equation for calculation the share of i-th fraction (α_{fi}) is given in the formula:

$$\alpha_{fi} = \left(\frac{m_{fi}}{\sum_{i=1}^{4} m_{fi}} \right) \cdot 100\% \tag{1}$$

Methodology for Preparing Biomass Mixtures

Compaction tests were carried out separately for each isolated fraction (samples A1, A2, A3, A4) and for several specially prepared mixtures. This was done to evaluate the parameters of the briquetting process, the durability of the briquettes produced from individual biomass fractions, and then the impact of the fractional composition on these parameters.

A summary of the percentage distribution of fractions in individual test series is presented in Table 1.

Table 1. Percentage share of fractions for briquetted mixtures.

No	Fraction Share (α_{fi}), %				Series Symbol
	f_1 (0 ÷ 1 mm)	f_1 (1 ÷ 4 mm)	f_1 (4 ÷ 8 mm)	f_1 (8 ÷ 16 mm)	
1.	100	0	0	0	A1
2.	0	100	0	0	A2
3.	0	0	100	0	A3
4.	0	0	0	100	A4
5.	50	25	25	0	B
6.	75	0	0	25	C
7.	75	0	25	0	D
8.	75	25	0	0	E

Source: own study.

According to the literature recommendations given by Wang et al. [40], in mixtures B, C, D, and E a high proportion of the finest fraction f_1 was used.

The briquetting process for the separated fractions was investigated at three temperatures—22 °C, 73 °C, and 103 °C—while for the mixtures, two temperature levels of 73 °C and 103 °C were used. These temperatures referred to the temperatures of biomass inside the compaction die.

2.2. Material Moisture Content

Moisture content of tested biomass, amounting to 10%, was determined in accordance with the EN13183-1:2004 standard [41]. Samples of 50 ± 0.5 g were weighed on the electronic scales RADWAG WTC 600 (RADWAG, Radom, Poland) with an accuracy of ±0.01 g and then after drying at 105 °C for 24 h the moisture content (*MC*, %) was determined using a formula:

$$MC = \frac{m_{bw} - m_{bd}}{m_{bw}} \cdot 100\%$$

(2)

where m_{bw} is a mass of wet sample (g), m_{bd} is a mass of dry sample (g).

2.3. The Agglomeration Process

The agglomeration process was carried out on a stand equipped with a universal testing machine type Veb Thüringer Industriewerk Rauenstein (TIRA, Germany) with the maximum force of 100 kN and closed die [37]. The obtained unit pressure was around 60 MPa and was a result of pistons' size and die diameters [42–46]. The internal diameter of the die was 45 mm, the height of the die was 300 mm, and the speed of piston displacement was 2 mm·s^{-1}. The volume of the compaction die was 477 cm^3, therefore the obtained samples had a such volume (Figure 1).

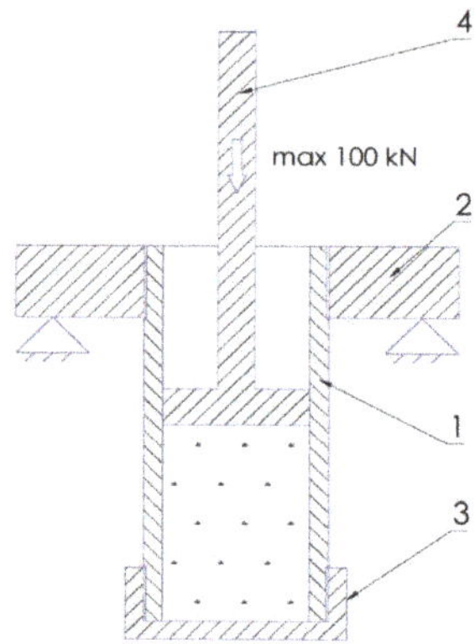

Figure 1. Schematic diagram of agglomeration stand (compaction die); 1—compaction die, 2—support, 3—bottom, 4—piston [37].

The temporary force occurring on the piston (±1 N) and the piston displacement (±0.01 mm) were recorded in the program HBM Catman v.2.1 (Hottinger Baldwin Messtechnik GmbH, Darmstad, Germany) with a frequency of 1 Hz. The maximum unit pressure of piston was 63 MPa. To maintain the temperature with an accuracy of ±1 °C micanite heaters connected to the controller EMKO ESM-3710 (EMKO Elektronik A.S., Bursa, Turkey) were mounted on the external surface of compaction die. A device TM2000 (Lutrom Electronic Enterprise Co LTD, Taiwan) with thermocouple type K were used to measure the temperature of the biomass inside the die with an accuracy of ±0.5 °C. The mass of a single dose of material was checked before each test on the laboratory scales WTC 600 (RADWAG, Radom, Poland) with an accuracy of ±0.01 g.

2.4. The Density of the Briquettes

The densities (ρ, kg·m^{-3}) of the briquettes, related to dry matter (DM), were determined using mass of the compacted sample, its moisture, and its volume at the maximum compaction pressure.

$$\rho_{dry} = \frac{m_p}{v}\cdot\left(1 - \frac{MC}{100}\right) \tag{3}$$

where: ρ_{dry} is a briquette density related to DM (kg·m^{-3}), m_p is a mass of the material with specific initial moisture content (kg), v is a briquette volume at maximum pressure of agglomeration (m^{-3}), MC is an initial moisture of compacted material (%).

2.5. Specific Work of Compaction

The total compaction work L_T was determined arithmetically on the basis of compaction force measurement results as a function of the piston stroke.

$$L_T = \int_{x_p}^{x_k} f(x)dx \approx \frac{x_k - x_p}{n}\left(\sum_{i=1}^{n-1} f(x_p + i\frac{x_k - x_p}{n}) + \frac{f(x_p) + f(x_k)}{2}\right) \tag{4}$$

The specific work of compaction was determined using the equation [47]:

$$L_u = \frac{L_T}{m_p} \tag{5}$$

where: L_T is a total compaction work (J), L_u is a specific work of compaction (J·g^{-1}), f is a function of compaction force to piston displacement, x_k, x_p are coordinates of the pistons' elementary displacement (m), n is the number of the pistons' elementary displacements, m_p is a mass of compacted material (g), and i is a dividing point counter.

2.6. Durability of Briquettes and Biomass Susceptibility to Compaction

The mechanical durability of agglomerates was determined according to the EN-ISO 17831-2:2016-02 standard [48]. The mass of tested samples was 2.0 ± 0.1 kg. The material was placed in the drum chamber with rotational speed of 21 rev·min^{-1} and 105 rotations were done [49,50]. Briquettes were separated from smaller particles and dust after the trial using a screen with openings of diameters equal to two thirds of a single briquette diameter. Then the sieved material was weighed on the electronic scales WLC 6/12/F1/R (RADWAG, Radom, Poland) with an accuracy of 0.1 g and the durability coefficient (Ψ) was calculated using the formula:

$$\Psi = \frac{m_{pt}}{m_p}\cdot 100 \tag{6}$$

where m_{pt} is a mass of briquettes after the durability test (g), m_p is a mass of briquettes before the durability test (g).

The value of susceptibility to compaction index k_c ((J·g^{-1})/(g·cm^{-3})) was determined using the equation:

$$k_c = \frac{L_u}{\rho_c - \rho_p} \tag{7}$$

where ρ_c, ρ_p are maximum and initial densities of compacted biomass, respectively (g·cm^{-3}).

2.7. Statistical Analysis

Obtained results were analyzed using the Statistica v.13 program [51]. All tests were carried out in five replications. The effects between dependent variables for compaction temperatures and material moisture contents were determined using a multivariate ANOVA analysis at the significance level $\alpha = 0.05$ and mean analysis was done.

3. Results

3.1. The Effects of the Temperature and Fractional Composition of Compacted Biomass on the Density of Briquettes and Specific Work of Compaction

The values of densities for briquettes produced at different parameters of the process were summarized in Table 1. Tests of mixtures with symbol A were carried out at three temperatures: 22, 73, and 103 °C. Based on the analysis of these results and taking into account the fact, that in each case the lowest density was obtained at room temperature (22 °C), testing of the rest of the mixtures was limited to two temperatures. For the majority of tested mixtures, the highest density of produced briquettes was obtained for the temperature of 103 °C. For mixture E, the density at a temperature of 103 °C was 1.11 kg·m^{-3}. Analyzing all of the briquettes' densities found a slight difference between the values of this parameter for temperatures 73 and 103 °C. This observation may lead to the conclusion that there is no purpose in increasing the temperature of the compaction process of shredded logging residues to 103 °C. The results of the statistical analysis for the density of briquettes (Table 2) indicate a significant impact of the fractional composition on their densities. However, both the process temperature and the interaction between the tested parameters had no significant effect on the densities of briquettes.

Table 2. Results of ANOVA analysis for factors influencing the durability coefficient (Ψ), specific work of compaction (L_c), susceptibility to compaction index (k_c) and briquette density (ρ_{dry}).

	df	SS	MS	F	*p*-Value
Briquette density ρ_{dry}					
MIX (A)	3	0.63	0.21	31.95	<0.0001
Temperature (B)	1	0.02	0.02	3.01	0.0867
Interaction (A × B)	10	0.06	0.01	0.94	0.5055
Error	80	0.53	0.01		
Specific work of compaction L_c					
MIX (A)	3	870.46	290.15	210.95	<0.0001
Temperature (B)	1	0.36	0.36	0.26	0.6120
Interaction (A × B)	10	158.92	15.89	11.55	<0.0001
Error	80	110.04	1.38		
Susceptibility to compaction index k_c					
MIX (A)	3	2897.95	965.98	105.94	<0.0001
Temperature (B)	1	36.30	36.30	3.98	0.0494
Interaction (A × B)	10	527.70	52.77	5.79	0.0002
Error	80	729.48	9.12		
Durability coefficient Ψ					
MIX (A)	3	0.14	0.05	14.28	0.0004
Temperature (B)	1	0.06	0.06	18.74	0.0001
Interaction (A × B)	3	0.20	0.07	20.58	<0.0001
Error	32	0.11	0.003		

The values of the specific work of compaction were very diverse (Table 3). For fine fractions (samples A1, A2, and A3) their values did not exceed 18 J·g^{-1}, regardless of the process temperature. The highest value of specific work of compaction was obtained during compaction of sample E at 103 °C, in which chips of fraction f_1 and f_2 were in a proportion of 75% to 25%, respectively. This effect may result from a very good packing of the finest biomass particles, wherein the addition (25%) of the f_2 fraction contributed to the plasticity of the compacted biomass. In the case of the four tested mixtures B, C, D, and E, in each case an increase in the specific work of compaction was observed due to the increase in the temperature. The increased temperature caused an increase in the plasticity of the biomass and a higher density (longer movement of the compaction piston for the same mass of biomass). Taking into account the level of test probability (Table 2), which is significantly lower than the significance level of $p = 0.05$, there was a strong relationship between the fractional composition and between the interaction (fractional composition, temperature) and specific work of compaction. There was no such relationship for the temperature of this process.

Table 3. The effects of interactions of temperature and fractional composition for mean values of briquettes densities (ρ_{dry}, kg·m^{-3}) related to the dry matter; ($F_{(2,\,24)} = 11.81$; $p = 0.0003$; SE = 2.75) and specific work of compaction (L_u, J·g^{-1}); ($F_{(2,\,24)} = 15.94$; $p < 0.0001$; SE = 0.67).

Series Symbol	Temp.	Briquettes Density	Specific Work of Compaction
	°C	kg·m^{-3}	J·g^{-1}
	22	1.04 ± 0.11	15.52 ± 0.82
A1	73	1.10 ± 0.12	11.61 ± 0.74
	103	1.11 ± 0.07	12.38 ± 0.66
	22	0.98 ± 0.05	17.15 ± 1.25
A2	73	1.04 ± 0.09	11.72 ± 0.97
	103	1.05 ± 0.08	15.80 ± 1.38
	22	0.84 ± 0.09	15.97 ± 1.19
A3	73	1.01 ± 0.08	14.17 ± 1.72
	103	1.04 ± 0.08	13.90 ± 1.27
	22	0.77 ± 0.08	28.26 ± 1.38
A4	73	0.80 ± 0.07	20.39 ± 1.75
	103	0.86 ± 0.06	20.06 ± 2.30
B	73	1.06 ± 0.03	29.40 ± 0.72
	103	1.14 ± 0.06	34.25 ± 0.53
C	73	0.90 ± 0.07	31.83 ± 0.91
	103	0.96 ± 0.09	33.62 ± 0.83
D	73	1.02 ± 0.08	29.83 ± 0.84
	103	1.06 ± 0.10	34.13 ± 0.94
E	73	1.14 ± 0.09	28.14 ± 0.78
	103	1.11 ± 0.06	34.55 ± 0.79

3.2. The Effects of the Temperature and Fractional Composition of Compacted Biomass on the Susceptibility to Compaction and Durability of Briquettes

Similarly, as in the case of the results of specific work of compaction, significant differences in the values of susceptibility to compaction were observed (Table 4). Significantly lower values were obtained for the three smallest fractions (from about 13.2 to about 22.5 (J·g^{-1})/(g·cm^{-3})). Analyzing the results summarized in Table 4, it was noted that the share of the thickest fraction f_4 increased the susceptibility to compaction index. For compaction of the A4 series (100% fraction of f_4), the highest value of the discussed parameter (42.40 (J·g^{-1})/(g·cm^{-3})) was obtained at a process temperature of 22 °C. For the mixture C with the fraction f_4 of 25% (75% of fraction f_1), the susceptibility to compaction at a process temperature of 73 °C was 39.62 (J·g^{-1})/(g·cm^{-3}). In most of the studied cases, increasing the process temperature resulted in a reduction in susceptibility to compaction. The exception was the compaction of samples from the E series, where the increase in temperature was observed together with an increase in susceptibility to compaction. The results of statistical analysis indicated a significant impact of the process temperature on the biomass susceptibility to compaction. In each case, the test probability level (Table 2) was lower than the significance level of $p = 0.05$.

The durability of briquettes obtained from the separated fractions of shredded logging residues is presented in Table 4. Some briquettes were automatically broken down into smaller parts, which is why Table 4 contains the results of durability tests only for mixtures of fractions. Satisfactory durability was not obtained in any of the analyzed cases. The briquettes of B series (50% of fraction f_1, 25% of fraction f_2, 25% of fraction f_3) were characterized by the highest durability of 68.1% and 72.5% for temperatures 73 °C and 103 °C, respectively. In three of the four tested mixtures, the increase in temperature improved the durability of the obtained briquettes.

Table 4. The effects of interactions of temperature fractional composition for mean values of susceptibility to compaction index (k_c, $(\text{J·g}^{-1})/(\text{g·cm}^{-3})$); $F_{(2, 24)} = 4.33$; $p = 0.0248$; SE = 1.04) and durability coefficient (Ψ, %); ($F_{(2, 24)} = 7.34$; $p = 0.0033$, SE = 0.01).

Series Symbol	Temp.	Susceptibility to Compaction Index	Durability Coefficient
	°C	$(\text{J·g}^{-1})/(\text{g·cm}^{-3})$	%
A1	22	18.82 ± 1.86	0.0
	73	13.20 ± 2.20	0.0
	103	13.92 ± 1.78	0.0
A2	22	20.49 ± 1.65	0.0
	73	13.11 ± 0.83	0.0
	103	17.63 ± 2.40	0.0
A3	22	22.54 ± 4.01	0.0
	73	15.92 ± 2.15	0.0
	103	15.22 ± 1.31	0.0
A4	22	42.40 ± 4.57	0.0
	73	29.26 ± 4.72	0.0
	103	26.34 ± 2.59	0.0
B	73	26.93 ± 0.97	68.1 ± 6.2
	103	24.17 ± 1.93	72.5 ± 3.0
C	73	39.62 ± 6.22	42.0 ± 2.9
	103	30.12 ± 3.36	67.4 ± 5.2
D	73	29.83 ± 3.27	64.2 ± 5.9
	103	26.84 ± 2.70	50.9 ± 5.3
E	73	23.58 ± 2.79	51.1 ± 1.0
	103	26.44 ± 3.02	66.1 ± 2.7

This effect was caused by liquefaction of resinous substances and activation of natural binders contained in biomass, which after solidification constituted a natural binder in the compacted material. Only in the case of samples from the D series did an increase in temperature cause a reduction in the durability of briquettes. This is due to the fact that the 4–8 mm fraction, which constituted 25% of this mixture, contained mostly fragments of bark and needles, i.e., matter with worse binding properties (lower content of resin substances and hemicellulose). As in the case of testing the susceptibility to compaction, also in the case of durability, the results of statistical analysis indicated a significant effect of the process temperature on the biomass susceptibility to compaction. In each case, the probability level of testing (Table 2) was significantly lower than the significance level of $p = 0.05$.

4. Discussion

Analysis of changes in the values of the examined density, specific work of compaction, durability of briquettes, and susceptibility to compaction showed that the fractional composition of compacted biomass had a major impact on these parameters. The highest density was obtained for mixtures with 75% of the finest fraction (0–1 mm) with an addition of fractions (4–8 mm) or (8–16 mm). The data also showed that the lowest density was obtained for briquettes made of 100% from the thickest fraction (8–16 mm). The highest densities were obtained for the mixture E (75% of fraction 0–1 mm and 25% of fraction 1–4 mm) for both temperature values. These values are consistent with the requirements of PN-EN ISO 17225-3 standard [52], which states that for briquettes made from logging residues the density should be greater than 0.9 g·cm^{-3}. Statistical analysis of results did not show a significant effect of either temperature or interaction between temperature and fractional composition on the density of briquettes. In most of the analyzed cases, increase in the temperature caused an increase in the density. As Chen and Kuo [53] report, higher temperatures allow for obtaining higher density of briquettes. Similar trends were emphasized in works by Gürdil and Demirel [29], as well as Kaliyan and Morey [25,30], Taulbee et al. [31], and Tumuluru et al. [32,33]. Attention is drawn, however, to the fact that increasing the temperature to 103 °C does not significantly increase the density. Considering the higher energy expenditure, it should be indicated that agglomeration at a higher temperature is less economically viable.

The obtained maximum density of individual briquettes did not exceed 1.14 g·cm^{-3}, which is the value corresponding to the density of pellets. As stated by Tumuluru et al. [33], due to the limited pressure, the briquettes have unit densities smaller than 1 g·cm^{-3}. High density of agglomerates is a required phenomenon from the point of view of transport and storage, and as recommended by Kaur et al. [54] is also beneficial for the combustion process, because it extends burning time.

The analysis of the data presented in Table 2 indicated that the specific work of compaction was strongly dependent on the fractional composition and the interaction between the fractional composition and temperature. On the other hand, the process temperature itself had no significant effect on the specific work of compaction. A detailed analysis of the test results (Table 3) showed that for all examined mixtures B, C, D, and E an increase in temperature resulted in an increase in the specific work of compaction. This is in line with the trend for other biological materials. The results were not clear and no similar relationship was noticeable in the case of compaction of separated fractions.

Durability is an important factor in proving the quality of briquettes. Similarly to Taulbee et al. [31], we found that increase in compaction temperature causes an increase in the strength of briquettes. In the studied range, the temperature had a significant influence on the durability of briquettes made of shredded logging residues. The used woody material is based on cellulose, lignin, hemicellulose, and resin, which at high temperature act as a binder [31], which improves the durability of briquettes. Also, according to Lisowski, the increase in temperature causes plasticizing of the particles and activation of natural binders in the material [35], which in turn leads to improved product durability. Unfortunately, in all cases the mechanical durability coefficient of tested briquettes ranging 42–72.5% was unacceptable and lower than 80–98% was obtained for briquettes made of wood sawdust, shredded plants, or crushed cones [55–60]. The highest value of durability coefficient in presented research was obtained for mixture B (50% of fraction 0–1 mm, 25% of fraction 1–4 mm, and 25% of fraction 4–8 mm) compacted at 103 °C. Size of particles and share of specific fractions also affected the low values of durability. According to Taulbee et al. [31] and Tumuluru et al. [32,33] a greater contact surface and more durable intermolecular connections are found for smaller particles. The obtained results correspond to the results of Ndindeng et al. [61], who report that briquette durability and density increase with decrease in agglomerated particle sizes. Moreover, according to Chou et al. [62] and Ryu et al. [63] smaller particles can result in a more dense structure. The volume of a single compacted sample could also have an adverse effect on the durability of obtained briquettes. It should also be stated that, according to many literature sources [64], the height of the compaction die has a significant impact on the biomass agglomeration process. A high volume of sample (height of the die) had a negative effect on the pressure distribution in the volume of the produced briquette.

5. Conclusions

Studies have confirmed the significant impact of the fractional composition of compacted biomass on its susceptibility to process parameters and the quality of the final product. Statistical analysis confirmed that the density of the briquette, its durability, specific work of compaction, and the susceptibility of the tested biomass to compaction strongly depend on the particle size of compacted biomass.

The obtained values of densities ranged from 0.768 to 1.14 g·cm^{-3} were consistent with those obtained by many researchers for various types of compacted material and the requirements contained in relevant standards.

The most favorable, from the point of view of density and durability of briquettes, was the temperature of 73 °C. Unfortunately, also at this temperature a satisfactory durability coefficient was not obtained.

Specific work of compaction depends on the process temperature. An increase in temperature to 73 °C increased this parameter by about 40%.

Increase in the process temperature up to 73 °C results in an increase in the susceptibility to compaction index.

Taking into account the great potential of forest biomass, it is crucial to carry out more research on this feedstock. In particular, tests should refer to the impact of the volume of the sample in single compaction (height of the die) on the durability and density of briquette.

Author Contributions: Conceived and designed the experiments: T.N., A.G. and K.R.; performed the experiments: T.N., A.G. and K.R.; analyzed the data: T.N., A.G. and M.D.; writing—original draft preparation, review and editing: T.N., A.G. and M.D.; supervision: T.N. All authors have read and agreed to the published version of the manuscript.

Funding: This research received no external funding.

Conflicts of Interest: The authors declare no conflict of interest.

References

1. Davis, S.J.; Caldeira, K. Consumption-based accounting of CO_2 emissions. *Proc. Natl. Acad. Sci. USA* **2010**, *107*, 5687–5692. [CrossRef] [PubMed]
2. Veum, K.; Bauknecht, D. How to reach the EU renewables target by 2030? An analysis of the governance framework. *Energy Policy* **2019**, *127*, 299–307. [CrossRef]
3. Tucki, K.; Mruk, R.; Orynycz, O.; Wasiak, A.; Botwińska, K.; Gola, A. Simulation of the Operation of a Spark Ignition Engine Fueled with Various Biofuels and Its Contribution to Technology Management. *Sustainability* **2019**, *11*, 2799. [CrossRef]
4. Jacobson, M.Z. Review of solutions to global warming, air pollution, and energy security. *Energy Environ. Sci.* **2009**, *2*, 148–173. [CrossRef]
5. Tucki, K.; Mruk, R.; Orynycz, O.; Wasiak, A.; Świć, A. Thermodynamic Fundamentals for Fuel Production Management. *Sustainability* **2019**, *11*, 4449. [CrossRef]
6. Abbas, D.; Current, D.; Phillips, M.; Rossman, R.; Hoganson, H.; Brooks, K.N. Guidelines for harvesting forest biomass for energy: A synthesis of environmental considerations. *Biomass Bioenergy* **2011**, *35*, 4538–4546. [CrossRef]
7. Aherne, J.; Posch, M.; Forsius, M.; Lehtonen, A.; Härkönen, K. Impacts of forest biomass removal on soil nutrient status under climate change: A catchment-based modelling study for Finland. *Biogeochemistry* **2012**, *107*, 471–488. [CrossRef]
8. Eggers, J.; Melin, Y.; Lundström, J.; Bergström, D.; Öhman, K. Management Strategies for Wood Fuel Harvesting—Trade-Offs with Biodiversity and Forest Ecosystem Services. *Sustainability* **2020**, *12*, 4089. [CrossRef]
9. Yazan, D.M.; van Duren, I.; Mes, M.; Kersten, S.; Clancy, J.; Zijm, H. Design of sustainable second-generation biomass supply chains. *Biomass Bioenergy* **2016**, *94*, 173–186. [CrossRef]
10. Koirala, A.; Kizha, A.R.; De Hoop, C.F.; Roth, B.E.; Han, H.-S.; Hiesl, P.; Abbas, D.; Gautam, S.; Baral, S.; Bick, S.; et al. Annotated Bibliography of the Global Literature on the Secondary Transportation of Raw and Comminuted Forest Products (2000–2015). *Forests* **2018**, *9*, 415. [CrossRef]
11. Sahoo, K.; Bilek, E.; Bergman, R.; Mani, S. Techno-economic analysis of producing solid biofuels and biochar from forest residues using portable systems. *Appl. Energy* **2019**, *235*, 578–590. [CrossRef]
12. Moskalik, T.; Sadowski, J.; Zastocki, D. Some technological and economic aspects of logging residues bundling. *Sylwan* **2016**, *160*, 31–39.
13. Moskalik, T.; Sadowski, J.; Sarzyński, W.; Zastocki, D. Efficiency of slash bundling in mature coniferous stands. *Sci. Res. Essays* **2013**, *8*, 1478–1486. [CrossRef]
14. Moskalik, T.; Gendek, A. Production of Chips from Logging Residues and Their Quality for Energy: A Review of European Literature. *Forests* **2019**, *10*, 262. [CrossRef]
15. Hakkila, P.; Parikka, M. Fuel resources from the forest. In *Bioenergy from Sustainable Forestry: Guiding Principles and Practice*; Richardson, J., Björheden, R., Hakkila, P., Lowe, A.T., Smith, C.T., Eds.; Springer: Dordrecht, The Netherlands, 2002; pp. 19–48. [CrossRef]
16. Gendek, A.; Malaťák, J.; Velebil, J. Effect of harvest method and composition of wood chips on their caloric value and ash content. *Sylwan* **2018**, *162*, 248–257.

17. Asikainen, A.; Pulkkinen, P. Comminution of Logging Residues with Evolution 910R chipper, MOHA chipper truck, and Morbark 1200 tub grinder. *Int. J. For. Eng.* **1998**, *9*, 47–53.

18. Spinelli, R.; Hartsough, B.R.; Magagnotti, N. Testing Mobile Chippers for Chip Size Distribution. *Int. J. For. Eng.* **2005**, *16*, 29–36. [CrossRef]

19. Nati, C.; Spinelli, R.; Fabbri, P. Wood chips size distribution in relation to blade wear and screen use. *Biomass Bioenergy* **2010**, *34*, 583–587. [CrossRef]

20. Friedl, A.; Padouvas, E.; Rotter, H.; Varmuza, K. Prediction of heating values of biomass fuel from elemental composition. *Anal. Chim. Acta* **2005**, *544*, 191–198. [CrossRef]

21. Barontini, M.; Scarfone, A.; Spinelli, R.; Gallucci, F.; Santangelo, E.; Acampora, A.; Jirjis, R.; Civitarese, V.; Pari, L. Storage dynamics and fuel quality of poplar chips. *Biomass Bioenergy* **2014**, *62*, 17–25. [CrossRef]

22. Gendek, A.; Nawrocka, A. Effect of chipper knives sharpening on the forest chips quality. *Ann. Wars. Univ. Life Sci.-SGGW Agric.* **2014**, *64*, 97–107.

23. Lisowski, A.; Świętochowski, A. Mechanical durability of pellets and briquettes made from a miscanthus mixture without and with the separation of long particles. *Pol. Soc. Agric. Eng.* **2014**, *1*, 93–100.

24. Ishii, K.; Furuichi, T. Influence of moisture content, particle size and forming temperature on productivity and quality of rice straw pellets. *Waste Manag.* **2014**, *34*, 2621–2626. [CrossRef] [PubMed]

25. Kaliyan, N.; Morey, R.V. Factors affecting strength and durability of densified biomass products. *Biomass Bioenergy* **2009**, *33*, 337–359. [CrossRef]

26. Križan, P.; Šooš, L.; Vukelić, D. A study of impact technological parametres on the briquetting process. *Work. Living Environ. Prot.* **2009**, *6*, 39–47.

27. Lee, S.H.; Lee, M.; Yoon, W.J.; Kim, Y. Frost growth characteristics of spirally-coiled circular fin-tube heat exchangers under frosting conditions. *Int. J. Heat Mass Transf.* **2013**, *64*, 1–9. [CrossRef]

28. Stelte, W.; Holm, J.K.; Sanadi, A.R.; Barsberg, S.; Ahrenfeldt, J.; Henriksen, U.B. A study of bonding and failure mechanisms in fuel pellets from different biomass resources. *Biomass Bioenergy* **2011**, *35*, 910–918. [CrossRef]

29. Gürdil, G.A.K.; Demirel, B. Effect of Particle Size on Surface Smoothness of Bio-Briquettes Produced from Agricultural Residues. *Manuf. Technol.* **2018**, *18*, 742–747. [CrossRef]

30. Kaliyan, N.; Morey, R.V. Strategies to improve durability of switchgrass briquettes. *Trans. ASABE* **2009**, *52*, 1943–1953. [CrossRef]

31. Taulbee, D.; Patil, D.P.; Honaker, R.Q.; Parekh, B.K. Briquetting of Coal Fines and Sawdust Part I: Binder and Briquetting-Parameters Evaluations. *Int. J. Coal Prep. Util.* **2009**, *29*, 1–22. [CrossRef]

32. Tumuluru, J.S.; Wright, C.; Kenney, K.; Hess, R. A technical review on biomass processing: Densification, preprocessing, modeling and optimization. In Proceedings of the American Society of Agricultural and Biological Engineers Annual International Meeting 2010 (ASABE 2010), Pittsburgh, PA, USA, 20–23 June 2010; Volume 6, pp. 20–23.

33. Tumuluru, J.S.; Wright, C.T.; Hess, J.R.; Kenney, K.L. A review of biomass densification systems to develop uniform feedstock commodities for bioenergy application. *Biofuels Bioprod. Biorefining* **2011**, *5*, 683–707. [CrossRef]

34. Whittaker, C.; Shield, I. Factors affecting wood, energy grass and straw pellet durability—A review. *Renew. Sustain. Energy Rev.* **2017**, *71*, 1–11. [CrossRef]

35. Lisowski, A.; Dąbrowska-Salwin, M.; Ostrowska-Ligęza, E.; Nawrocka, A.; Stasiak, M.; Świętochowski, A.; Klonowski, J.; Sypuła, M.; Lisowska, B. Effects of the biomass moisture content and pelleting temperature on the pressure-induced agglomeration process. *Biomass Bioenergy* **2017**, *107*, 376–383. [CrossRef]

36. Nurek, T.; Gendek, A.; Roman, K. Forest Residues as a Renewable Source of Energy: Elemental Composition and Physical Properties. *BioResources* **2019**, *14*, 6–20. [CrossRef]

37. Nurek, T.; Gendek, A.; Roman, K.; Dąbrowska, M. The effect of temperature and moisture on the chosen parameters of briquettes made of shredded logging residues. *Biomass Bioenergy* **2019**, *130*, 105368. [CrossRef]

38. International Organization for Standardization. *Solid Biofuels—Determination of Particle Size Distribution for Uncompressed Fuels—Part 1: Oscillating Screen Method Using Sieves with Apertures of 3,15 mm and Above*; ISO 17827-1:2016; International Organization for Standardization: Geneva, Switzerland, 2016.

39. International Organization for Standardization. *Solid Biofuels—Determination of Particle Size Distribution for Uncompressed Fuels—Part 2: Vibrating Screen Method Using Sieves with Aperture of 3,15 mm and Below*; ISO 17827-2:2016; International Organization for Standardization: Geneva, Switzerland, 2016.

40. Wang, Y.; Wu, K.; Sun, Y. Effects of raw material particle size on the briquetting process of rice straw. *J. Energy Inst.* **2018**, *91*, 153–162. [CrossRef]

41. European Committee for Standardization. *Moisture Content of a Piece of Sawn Timber—Part 1: Determination by Oven Dry Method*; EN 13183-1:2004; European Committee for Standardization: Brussels, Belgium, 2004.

42. Hejft, R. Manufacturing of briquettes from plant waste in a worm tool-in-use system. *Inż. Rol.* **2006**, *5*, 231–238.

43. Kers, J.; Kulu, P.; Aruniit, A.; Laurmaa, V.; Križan, P.; Šooš, L.; Kask, Ü. Determination of physical, mechanical and burning characteristics of polymeric waste material briquettes. *Est. J. Eng.* **2010**, *16*, 307–316. [CrossRef]

44. Frączek, J.; Mudryk, K.; Wróbel, M. Energy expenditures in the process involving Salix viminalis L. Willow briquetting. *Inż. Rol.* **2010**, *3*, 45–52.

45. Sergeev, P.V.; Beletskiy, V.S. Briquetting the carbon phase from the sludge ponds at the Anzhersk deposit. *Coke Chem.* **2013**, *56*, 282–285. [CrossRef]

46. Kulig, R.; Skonecki, S.; Gawłowski, S.; Zdybel, A.; Łysiak, G. The effect of pressure on efficiency of chosen soft wood sawdust. *Acta Sci. Pol. Tech. Agrar.* **2013**, *12*, 31–40.

47. Laskowski, J.; Skonecki, S. Effect of chamber parameters and material weight on densification of lupine seeds. *Inż. Rol.* **2005**, *9*, 101–108.

48. Polish Committee for Standardization. *Solid Biofuels—Methods for the Determination of Mechanical Durability of Pellets and Briquettes—Part 2: Briquettes*; EN-ISO 17831-2:2016-02; Polish Committee for Standardization: Warsaw, Poland, 2016.

49. Hebda, T.; Złobecki, A. Influence of straw moisture on kinetics durability of briquettes. *Inż. Rol.* **2011**, *6*, 45–52.

50. Temmerman, M.; Rabier, F.; Daugbjerg Jensen, P.; Hartmann, H.; Böhm, T. Comparative study of durability test methods for pellets and briquettes. *Biomass Bioenergy* **2006**, *30*, 964–972. [CrossRef]

51. Dell Inc. *Dell Statistica (Data Analysis Software System)*; Version 13; Dell Inc.: Round Rock, TX, USA, 2016; Available online: software.dell.com (accessed on 19 June 2019).

52. International Organization for Standardization. *Solid Biofuels—Fuel Specifications and Classes—Part 3: Graded Wood Briquettes*; ISO 17225-3:2014; International Organization for Standardization: Geneva, Switzerland, 2014.

53. Chen, W.-H.; Kuo, P.-C. A study on torrefaction of various biomass materials and its impact on lignocellulosic structure simulated by a thermogravimetry. *Energy* **2010**, *35*, 2580–2586. [CrossRef]

54. Kaur, A.; Roy, M.; Kundu, K. Densification of biomass by briquetting: A review. *Int. J. Recent Sci. Res.* **2017**, *8*, 20561–20568. [CrossRef]

55. Gendek, A.; Aniszewska, M.; Malaťák, J.; Velebil, J. Evaluation of selected physical and mechanical properties of briquettes produced from cones of three coniferous tree species. *Biomass Bioenergy* **2018**, *117*, 173–179. [CrossRef]

56. Niedziółka, I.; Szymanek, M.; Zuchniarz, A. Durability evaluation for briquettes made of fodder corn phytomass. *Inż. Rol.* **2008**, *9*, 235–240.

57. Niedziółka, I.; Szpryngiel, M.; Kraszkiewicz, A.; Kachel-Jakubowska, M. Assessments of briquetting efficiency and briquettes quality produced out of selected plant raw materials. *Inż. Rol.* **2011**, *6*, 149–155.

58. Fiszer, A. Influence of straw humidity and temperature of briquetting processon the quality of agglomerate. *J. Res. Appl. Agric. Eng.* **2009**, *54*, 68–70.

59. Borkowski, Ł.; Gendek, A. Określenie trwałości brykietów wytworzonych z materiału pochodzenia rolniczego i leśnego (Determining the durability of briquettes produced from agricultural and forest materials). In Proceedings of the XXI Międzynarodowa Konferencja Naukowa Studentów: Współczesne Aspekty inżynierii Produkcji, Warszawa, Poland, 23 May 2012; pp. 19–26.

60. Mudryk, K. Quality assessment for briquettes made of biomass from maple [Acer negundo L.] and black locust [Robinia pseudoacacia L.]. *Agric. Eng.* **2011**, *7*, 115–121.

61. Ndindeng, S.A.; Mbassi, J.E.G.; Mbacham, W.F.; Manful, J.; Graham-Acquaah, S.; Moreira, J.; Dossou, J.; Futakuchi, K. Quality optimization in briquettes made from rice milling by-products. *Energy Sustain. Dev.* **2015**, *29*, 24–31. [CrossRef]

62. Chou, C.-S.; Lin, S.-H.; Peng, C.-C.; Lu, W.-C. The optimum conditions for preparing solid fuel briquette of rice straw by a piston-mold process using the Taguchi method. *Fuel Process. Technol.* **2009**, *90*, 1041–1046. [CrossRef]

63. Ryu, C.; Yang, Y.B.; Khor, A.; Yates, N.E.; Sharifi, V.N.; Swithenbank, J. Effect of fuel properties on biomass combustion: Part I. Experiments—Fuel type, equivalence ratio and particle size. *Fuel* **2006**, *85*, 1039–1046. [CrossRef]

64. Lisowski, A.; Pajor, M.; Świętochowski, A.; Dąbrowska, M.; Klonowski, J.; Mieszkalski, L.; Ekielski, A.; Stasiak, M.; Piątek, M. Effects of moisture content, temperature, and die thickness on the compaction process, and the density and strength of walnut shell pellets. *Renew. Energy* **2019**, *141*, 770–781. [CrossRef]

MDPI
St. Alban-Anlage 66
4052 Basel
Switzerland
Tel. +41 61 683 77 34
Fax +41 61 302 89 18
www.mdpi.com

Sustainability Editorial Office
E-mail: sustainability@mdpi.com
www.mdpi.com/journal/sustainability

9 783036 520513